The Scientific Revolution

Blackwell Essential Readings in History

This series comprises concise collections of key articles on important historical topics. Designed as a complement to standard survey histories, the volumes are intended to help introduce students to the range of scholarly debate in a subject area. Each collection includes a general introduction and brief contextual headnotes to each article, offering a coherent, critical framework for study.

Published

The Scientific Revolution

The Essential Readings

Edited by Marcus Hellyer

Blackwell
Publishing

Editorial material and organization © 2003 by Blackwell Publishing Ltd

350 Main Street, Malden, MA 02148-5018, USA
9600 Garsington Road, Oxford OX4 2DQ, UK
550 Swanston Street, Carlton South, Melbourne, Victoria 3053, Australia
Kurfürstendamm 57, 10707 Berlin, Germany

First published 2003 by Blackwell Publishing Ltd

Library of Congress Cataloging-in-Publication Data
The scientific revolution: the essential readings / edited by Marcus Hellyer.
    p.   cm. – (Blackwell essential readings in history)
   Includes bibliographical references and index.
   ISBN 0-631-23629-5 (alk. paper) – ISBN 0-631-23630-9 (pbk. : alk. paper)
   1. Science–Europe–History–16th century.   2. Science–Europe–History–17th century.
I. Hellyer, Marcus. II. Series.

   Q125.2 .S37 2003
   509.4′09′031 – dc21

                    2002034220

A catalogue record for this title is available from the British Library.

Set in 10 on 12pt Photina
by SNP Best-set Typesetter Ltd., Hong Kong
Printed and bound in the United Kingdom
by MPG Books Ltd, Bodmin, Cornwall

For further information on
Blackwell Publishing, visit our website:
http://www.blackwellpublishing.com

Contents

Acknowledgments

Original Sources of Articles

The reader should note that several pieces have been slightly abridged, particularly footnotes.

The editor and publishers wish to thank the following for permission to use copyright material:

R. Hooykaas, "The Rise of Modern Science: When and Why?" *British Journal for the History of Science*, 20 (1987), pp. 453–73.

Robert S. Westman, "The Copernicans and the Churches," in *God and Nature: Historical Essays on the Encounter between Christianity and Science*, eds. David C. Lindberg and Ronald L. Numbers (Berkeley: University of California Press, 1986), pp. 76–113. Copyright © 1986 The Regents of the University of California Press, 1986.

Peter Dear, "A Mechanical Microcosm: Bodily Passions, Good Manners, and Cartesian Mechanism," in *Science Incarnate: Historical Embodiments of Natural Knowledge*, ed. Christopher Lawrence and Steven Shapin (Chicago: University of Chicago Press, 1998), pp. 51–82.

Steven Shapin, "Pump and Circumstance: Robert Boyle's Literary Technology," *Social Studies of Science*, (14) 1984, pp. 481–520.

William B. Ashworth, Jr., "Natural History and the Emblematic World View," in *Reappraisals of the Scientific Revolution*, eds. David C. Lindberg and Robert S. Westman (Cambridge: Cambridge U. P., 1990), pp. 303–33. © Cambridge University Press.

Allen G. Debus, "The Chemical Philosophy and the Scientific Revolution," in *Revolutions in Science: Their Meaning and Relevance*, ed. William R. Shea (Canton, Mass.: Science History Publications, 1988), pp. 27–48.

I. Bernard Cohen, *Revolution in Science*, (Cambridge, Mass.: Belknap Press of Harvard University Press, 1985). © 1985 by the President and Fellows of Harvard College.

Margaret C. Jacob, *Scientific Culture and the Making of the Industrial West*, (Oxford: Oxford U. P., 1997).

Andrew Cunningham and Perry Williams, "De-centring the 'big picture': The Origins of Modern Science and the modern origins of science," *British Journal for the History of Science*, 26 (1993), pp. 407–32.

Every effort has been made to trace copyright holders and to obtain their permission for the use of copyright material. The authors and publishers will gladly receive any information enabling them to rectify any error or omission in subsequent editions.

Editor's Introduction: What was the Scientific Revolution?

Marcus Hellyer

Science is one of the most powerful forces in modern society, having enormous social, cultural, economic, and military influence. The science generating the technologies of the digital revolution is a crucial driver of modern economies. Chemistry and physics transformed the nature of warfare in the twentieth century. Legal cases, both civil and criminal, are often settled – or obscured – by the testimony of scientists as expert witnesses. People suffering from terminal disease wait in hope that scientists can develop a cure. In short, science pervades modern western societies, and through the processes of globalization it is coming to play a comparable role in many non-western societies.

This has not always been the case; for most of history there was no enterprise even remotely similar to our science. Thus, science has a history, although its history is not just a parade of changing, improving theories about particular natural phenomena, but a real history of human activity, of how and why we have investigated nature.

This history has been episodic. The investigation of nature has been relatively static in its contents, methods, and purposes at some times, but has changed rapidly at others. One era of fundamental change has come to be known as the Scientific Revolution. As with all revolutions, historians argue endlessly over its exact nature and periodization. Most agree that it occurred during the early modern period, from the early sixteenth century to the late seventeenth, although good cases can be made to extend it at either end. Historians are, however, virtually unanimous in regarding the Scientific Revolution as a European phenomena, occurring primarily in the western European regions of Italy, France, the Netherlands, Great Britain, and Germany.

Yet once we turn to the fundamental question, "What was the Scientific Revolution about?" we run into a lively debate. The answers that historians have provided to this question show why this is one of the most vibrant fields of the history of science and of history overall. The goal of this volume is to provide recent examples of the answers historians have given to this question. First this introduction will briefly outline historians' approaches to the Scientific Revolution for different approaches will give very different answers.[1]

The generation of scholars who introduced the history of science to the universities of Europe and North America in the decades either side of the Second World War created a coherent and compelling account of the Scientific Revolution. They focused primarily on the great advances in knowledge about natural phenomena and on theories in particular. This narrative primarily recounted the development of astronomy and mechanics and usually went something like this: it begins with Copernicus's (1473–1543) publication of his heliocentric cosmology in *De Revolutionibus Orbium Coelestium* (1543), in which he claimed the earth both rotated on its axis and revolved around the Sun. Copernicus was not seeking to overthrow ancient astronomy, but to restore perfect circular motion to the heavens. The story then moves to the late sixteenth century when a Danish nobleman Tycho Brahe (1546–1601) carried out astronomical observations of unprecedented accuracy and thoroughness on his island of Hven. Tycho also determined that comets and nova are celestial not meteorological phenomena, undermining Aristotle's (384–322 BCE) teaching on the perfection and immutability of the heavens.

The next figure to appear is Johannes Kepler (1571–1630), a German astronomer who was convinced of the truth of the Copernican system because it meshed seamlessly with the five perfect solids so important to his Neoplatonic world view, as he described in his *Cosmographic Mystery* of 1596. Kepler worked tirelessly yet unsuccessfully to reconcile Tycho's data to perfect circular orbits. Eventually he concluded that planetary orbits must be elliptical, thus breaking the ancient "spell" of perfect circular motion in the heavens, a finding published in his *Astronomia Nova* of 1609. At this point Galileo (1564–1642) enters the stage. Pointing the newly invented telescope at the heavens, he discovered phenomena which he described in his celebrated *Sidereus Nuncius* of 1610 such as the mountains on the moon, which further undermined the perfection of the heavens, and the moons of Jupiter, which indicated the earth need not be the center of all celestial motions. Galileo also set to work developing a system of mechanics to replace Aristotle's, in order to explain how heavy bodies can fly through the cosmos as well as fall down on earth, the results of which he eventually published in his *Two New Sciences* of 1638.

1   H. Floris Cohen, *The Scientific Revolution: A Historiographical Inquiry* (Chicago: University of Chicago Press, 1994), provides a complete overview of historical writing on the subject.

The culmination of the story occurs with Isaac Newton (1642–1727). In his *Mathematical Principles of Natural Philosophy* of 1687, Newton developed a mechanics based on the theory of universal gravity and the three laws of motion that could account for all motions, whether on earth or in the heavens, thereby erecting a complete system of physics that finally replaced Aristotle. Along the way he developed the necessary mathematical tools, the calculus, contemporaneously and independently of its other inventor Gottfried Wilhelm Leibniz (1646–1716). This conveniently offers a finishing line for the cosmic relay race towards truth begun in 1543 with Copernicus's *De Revolutionibus*.[2]

This short outline risks parodying a story that is told skillfully by excellent scholars, and one that is not without historical accuracy and narrative coherence, even when fleshed out with numerous lesser figures.[3] But the sheer success of the narrative raised problems. For one, it did not find clear parallels in disciplines outside of astronomy and mechanics. If developments had to be equally transformative to qualify as a revolution, then revolutions were hard to come by in chemistry or the various fields of natural history and medicine.

Certainly there were major achievements in these fields. One could also tell a narrative of medical discovery, particularly in anatomy and physiology. This begins in 1543 (conveniently the same year as Copernicus's *De Revolutionibus*) with Andreas Vesalius's (1514–64) work in anatomy, *On the Structure of the Human Body*. Based on dissections he himself performed, it set new standards both for empirical investigation and scholarly publishing. Eventually his investigations forced Vesalius to reject basic elements of Galenic anatomy, such as a permeable septum between the left and right ventricles of the heart. The narrative progresses to William Harvey's (1578–1657) discovery of the circulation of the blood, described in his *On the Movement of the Heart and Blood* (1628), and Richard Lower's (1631–91) work on the function of the lungs and his experiments on blood transfusions. The development of microscopy, particularly by Antoni van Leeuwenhoek (1632–1723) and Robert Hooke (l635–1703), is another highpoint of the narrative, leading to Marcello Malpighi's (1628–94) discovery of the capillaries in 1661 with the microscope and Leeuwenhoek's correct interpretation of their function in 1683.

But these developments led neither to a fundamental change in medical practice nor to greater diagnostic or predictive accuracy rivaling that of the

2   A recent and very useful resource, Wilbur Applebaum (ed.), *Encyclopedia of the Scientific Revolution from Copernicus to Newton* (New York and London: Garland, 2000), also adopts this chronology.

3   Prominent examples of this approach include Alexandre Koyré, *The Astronomical Revolution: Copernicus–Kepler–Borelli* (New York; Dover, 1992 [1961]); Richard S. Westfall, *Force in Newton's Physics: the Science of Dynamics in the Seventeenth Century* (New York: American Elsevier, 1971).

astronomers. It was hard to assess this history as revolutionary. Similarly natural history has long been the neglected stepsister in the narrative of the Scientific Revolution, Although there was a vast increase in knowledge about plants and animals sent back to Europe by merchants, missionaries, and explorers from around the world, it did not map easily on to the model established in astronomy. Natural history, it was said, had to wait until the nineteenth century for Darwin.

In general, the various branches of what we now term physics remained a much easier sell as revolutionary. Certainly the development of experimental instruments such as Evangelista Torricelli's (1608–47) mercury tube barometer, invented in 1644, and Otto von Guericke's (1602–86) air-pump, devised in the 1650s and soon improved upon by Robert Boyle (1627–91) and Robert Hooke in England, was extraordinary. Such instruments allowed the investigation and identification of the various characteristics of atmospheric air, such as its weight, pressure, and elasticity, as well as its role in combustion, respiration, and the transmission of sound. Similar giant strides were made in the study of electricity in the eighteenth century. The work of Kepler, Newton, René Descartes (1596–1650), and many others in both theoretical and practical optics and the physiology of vision resulted in major advances such as the law of refraction, the determination of the composition of white light, and steadily superior telescopes and accurate observations.

Historians attempted to overcome this disunity of different disciplinary narratives by determining features that they all shared. One approach traced a trajectory of differentiation from the ancients. Most sixteenth-century practitioners did not consciously set out to create a revolution; they were trying to revive ancient scholarship. Later, as they became aware of its inadequacies, they sought to repair its gaps or inaccuracies, for example, by adding to Pliny's *Natural History* (first century CE) plants and animals from the New World (Hooykaas and Ashworth's essays discuss the impact of the New World), or by removing abhorrent constructions such as the equant from Ptolemy's (fl. second century CE) astronomical masterpiece, *The Almagest* (see Westman).

Gradually, this movement to restore and augment became a drive to overthrow and replace. Seventeenth-century practitioners were very conscious of the novelty of their enterprises. The term new (*nova* in Latin) occurs again and again in the titles of their works.[4] Outside of the university there was widespread hostility towards Aristotle in particular, the most influential of the ancient philosophers, whose works still dominated the universities' curriculum in the seventeenth century. Descartes hoped to replace Aristotle's natural

---

4   Johannes Kepler's *New Astronomy*, Galileo's *Two New Sciences*, and Francis Bacon's *New Atlantis* and *New Organum* are prominent examples. Otto von Guericke and Robert Boyle both wrote works on the air-pump entitled *New Experiments*.

philosophy in the curriculum with his own. Thus a trajectory that began in the Renaissance with restoring the ancients moved to repairing their gaps or errors, and finally to rejecting and replacing them.

Another approach that tried to unify various fields of natural knowledge focused on the metaphysical underpinnings of the Scientific Revolution.[5] One fundamental element of the new natural philosophy was termed the mechanical philosophy. Strongly influenced by actual machines, particularly clocks and automata, it sought to project a clockwork mechanism onto the entire universe and all its components. The mechanical philosophy held that all natural bodies consisted of particles of matter in motion and nothing else. All perception was the result of particles of matter acting upon the particles of matter that comprise our sensory organs. Descartes's version was perhaps the best known, but there were numerous competing schools of mechanical philosophy.[6] They disagreed on the number of kinds of particles (from one to three to many to infinite), on whether the particles are infinitely divisible or ultimately irreducible, on whether they move in a vacuum or a plenum, and on whether matter is completely inert and lacking in any active or spiritual element.

The mechanical philosophy did not just make ontological claims about the world, that is, the kinds of things that exist, but also epistemological claims, that is, what counted as real knowledge and explanation. Reacting to the substantial forms of the universities' Aristotelian philosophers and the endless occult similitudes, sympathies and other qualities of the natural magicians, the mechanical philosophers insisted that an explanation of physical processes had to be expressed in mechanical terms, in the direct contact of particles on other particles. This led to perhaps the greatest irony of the Scientific Revolution, namely that many of Newton's contemporaries did not regard his theory of gravity as being truly scientific since he refused to identify a mechanical cause of gravity. Yet while a narrative structure focusing on the mechanical philosophy can unify natural philosophy with psychology (as Dear's piece shows) and even physiology, many areas of investigation paid little or no attention to it, as Ashworth's essay argues.

A similar approach that seeks to provide a coherent narrative focuses on method, arguing that the fundamental novelty of the Scientific Revolution was how practitioners examined nature and made their claims. The problem is that the range of methods used in the investigation of nature in early

5   E. A. Burtt, *The Metaphysical Foundations of Modern Physical Science: A Historical and Critical Essay* (New York: Harcourt, Brace, 1925).
6   The literature on Descartes's natural philosophy is enormous, see William R. Shea, *The Magic of Numbers and Motion: The Scientific Career of René Descartes* (Canton, Mass.: Science History Publications, 1991); Daniel Garber, *Descartes' Metaphysical Physics* (Chicago: University of Chicago Press, 1992); Stephen Gaukroger et al. (eds), *Descartes' Natural Philosophy* (London and New York: Routledge, 2000).

modern Europe is as bewilderingly varied as the disciplines and their practitioners themselves. Descartes's rationalism that built from a priori first
principles was not the same as Newton's hypothetico-deductive method that
eschewed speculation. Furthermore, an explicit affirmation of a particular
method was often a declaration of adherence to the new philosophy, not necessarily an accurate statement about how one actually studied nature; Francis
Bacon's (1561–1626) admonition to proceed by the inductive collection of
individual facts was repeated approvingly by practitioners of all kinds, even
those whose own method differed radically from Bacon's.[7] Also, in the question of method, radical distinctions between medieval and early modern blur.
While Galileo's method, the regressus, has often been termed the modern
scientific method, it was largely derived from scholasticism, although his
insertion of mathematics and experiment into it was innovative. In addition,
Galileo's insistence that natural philosophy had to provide certain demonstrations stands in clear contrast to the insistence by Boyle and other experimentalists at the Royal Society on the provisional and hypothetical nature of
knowledge.[8]

As part of the process of rejecting the ancients outlined above, virtually all
fields of natural philosophy and natural history came to privilege experience
over authority. Hooykaas's piece stresses the importance of facts in the Scientific Revolution. But this emphasis certainly was not limited to the study of
nature, and indeed, as Ashworth argues, may have been imported from
humanistic disciplines such as antiquarianism. So a narrative that stresses the
privileging of facts over textual authority dilutes and dissipates the uniqueness
of natural inquiry in this period.

Other historians saw the revolutionary aspect as mathematical. In contrast
to earlier systems of natural knowledge which rigidly distinguished between
the domains of natural philosophy and mathematics, the Scientific Revolution
experienced a trajectory that integrated mathematics and natural philosophy,
culminating again in Newton's mathematical mechanics. This trajectory maps
very neatly on the astronomical and mechanical one outlined above. Copernicus famously wrote in his *De Revolutionibus* that he was writing mathematics
for mathematicians even though he was also speaking about the nature of the
heavens and not just their motions. Both Kepler and Galileo held that the Book
of Nature was written in the language of mathematics and only those versed

---

7   On Bacon see Julian Martin, *Francis Bacon, the State, and the Reform of Natural Philosophy*
(Cambridge: Cambridge University Press, 1992; Stephen Gaukroger, *Francis Bacon and the
Transformation of Early-Modern Natural Philosophy* (Cambridge: Cambridge University Press,
2001).
8   See William A. Wallace's numerous publications, particularly *Galileo and His Sources: The
Heritage of the Collegio Romano in Galileo's Science* (Princeton: Princeton University Press, 1984);
and his essays collected in *Galileo, the Jesuits and the Medieval Aristotle* (Aldershot: Variorum,
1991).

in geometry could comprehend it. The title of Newton's masterpiece proclaimed the complete integration of mathematics and natural philosophy. But again, how well does this mathematical revolution describe fields such as natural history, chemistry, or medicine?

These approaches were largely based on assumptions about science that can be termed "internalist." Again at risk of parody, this view held that science was primarily a cognitive activity quite distinct from society and culture. Science proceeded by observations and by the interaction and testing of theories. When society and culture did impinge upon science, it was as a foreign influence, quite different from science itself.[9]

But there was an alternative approach, one which saw the force driving science as coming from outside science itself, from the broader society. Marxist historians, for example, argued that the developments of the Scientific Revolution arose in response to the needs of early capitalism, in particular trade and navigation.[10] Certainly the great figures of Scientific Revolution were interested in solving practical problems. Many of them were intensely practical and technically gifted. But to reduce the Scientific Revolution to a quest for material and technological progress is a misrepresentation. The pursuit of truth and utility are hard to disentangle, even more so if we apply our standards of utility to the early modern period.

A more stimulating example of an externalist approach was provided in 1938 by Robert Merton, who suggested that certain social systems or ideologies provided values that were more conducive to science. Merton s particular case-study was seventeenth-century English Puritanism, which he claimed was in large part responsible for the flourishing of science in mid-seventeenth-century England.[11]

The divide between internalism and externalism can and has been overstated.[12] There was considerable overlap, and most historians from both camps accepted that there was a Scientific Revolution.[13] This "founding" generation of historians of science produced excellent studies that retain considerable merit. Two selections by members of that group (Reijer Hooykaas and

9   This is not to say that internalist historians did not see science as the product of Europe's unique culture and rationality.
10   J. D. Bernal, *Science in History. Vol. 2. The Scientific and Industrial Revolutions* (Cambridge, Mass.: The MIT Press, 1971 [1954]) presents a modern version of this approach.
11   Robert K. Merton, *Science, Technology & Society in Seventeenth-Century England* (New York: H. Fertig, 1970 [1938]).
12   For an overview of the debate see Steven Shapin, "Discipline and Bounding: The History and Sociology of Science as Seen Through the Externalism–Internalism Debate," *History of Science*, 30 (1992), pp. 333–69.
13   One should note that there is also a school of historians, mainly scholars of medieval science, who denied the revolutionary aspects of early modern science. Following the lead of Pierre Duhem in the early twentieth century, they argue that the origins of modern science lay in the Middle Ages and stress the continuities between medieval and early modern science.

I. Bernard Cohen) are included in this volume as illustrations their achievement. As a historical narrative, the Scientific Revolution perhaps reached its zenith in 1949 with Herbert Butterfield's famous comparison, in which he claimed it "outshines everything since the rise of Christianity and reduces the Renaissance and Reformation to the rank of mere episodes, mere internal displacements, within the system of medieval Christendom."[14]

Around the 1970s, however, historians of science adopted different approaches and consequently the questions they asked changed. There are several explanations for this development, one being the changing attitudes to science in general (discussed by Cunningham and Williams in this volume). Another shift was in the education of historians of science. Members of the founding generation had generally been trained as philosophers or scientists. Thus it is not surprising that they were interested in philosophical questions of method, metaphysics, and epistemology, or the internal histories of their scientific disciplines. Increasingly, however, scholars trained as historians entered the field. The historical profession had long since expanded from political and military history with its focus on "great men" to social and cultural history. They brought to the history of science the historian's concern for placing events in their wider cultural and social context.

Additionally, there was a fundamental shift in historians' understanding of the nature, production, and transmission of scientific knowledge. Many scholars from numerous disciplines contributed to this development, but among the most influential was Thomas Kuhn, who was, ironically, a philosophically inclined physicist.[15] Kuhn argued that scientists are trained and work within a paradigm that shapes their questions and methods, Most of what they do is simply normal science, solving problems that arise within the paradigm. Occasionally, as inexplicable anomalies accumulate, a new paradigm arises that can explain them. The new paradigm attracts its own adherents, mainly from the younger generations, thereby causing a revolution. Crucially, Kuhn insisted a scientist's decision to adhere to or abandon one paradigm could not be rationally explained, but depended on many cultural and social commitments. Later scholars, particularly sociologists, realized that Kuhn's work broke down the wall between science and society. In essence, all knowledge, even scientific knowledge, was social. Not only did members of scientific disciplines bring cultural commitments from other areas of their culture to their work, but the disciplines themselves were in effect societies which socialized their members. A vigorous sociological literature about science blossomed and offered historians

---

14 Herbert Butterfield, *The Origins of Modern Science, 1300–1800* (London: G. Bell, 1957 [1949]), p. vii.
15 Thomas S. Kuhn, *The Structure of Scientific Revolutions* (Chicago: University of Chicago Press, 1962).

the hope of overcoming the internalist/externalist debates.[16] Knowledge was a social product, and without social organizations, there could be nothing that counted as knowledge.

While we should not draw a line too sharply between internalist and externalist approaches, or between the older and more recent history of science, we can nonetheless identify significant results of the shift within the history of science. One of the most important has been an increased concern with the social role and self-conceptions of practitioners. They were not scientists as we know them, but were radically different in their views of themselves and their enterprise.[17] A significant element of early modern science was how these self-conceptions changed as new disciplinary boundaries and hierarchies arose. When, following his discovery of the moons of Jupiter, Galileo moved from being a university mathematician to the court philosopher to the Grand Duke of Tuscany, he did so not for a larger salary, but for the greater institutional freedom and higher disciplinary standing his new position offered. This standing, he hoped, would allow him to make claims that he could not as a mere mathematician.

Similarly, since science is an inherently social activity historians are devoting greater attention to the institutions in which it is practiced. Studies have moved beyond the flagship scientific societies, such as the Royal Philosophical Society of London (founded 1660) and the Paris Royal Academy of Sciences (1666), to include the regional learned societies that copied them on a local level. The princely courts provided opportunities for mathematicians, natural philosophers, and physicians and often directed their attention to particular problems – Galileo did not find the time he hoped for at court as he was constantly drawn into disputes in which he as court philosopher was obliged to participate.[18]

Historians now also count among the institutions that fostered science international "corporations" such as the Society of Jesus with its worldwide missions and colleges, and international trading companies like the Dutch East India Company that collected plant and animal specimens from around the

16   For example, Barry Barnes, *Scientific Knowledge and Sociological Theory* (London: Routledge and Kegan Paul, 1974). Jan Golinksi, *Making Natural Knowledge: Constructivism and the History of Science* (Cambridge: Cambridge University Press, 1998) provides an excellent discussion of the implications of the social and cultural understanding of science for the history of science.

17   One of the most influential early works on this subject was Robert S. Westman, "The Astronomer's Role in the Sixteenth Century: A Preliminary Study," *History of Science*, 18 (1980), pp. 105–47.

18   See the essays in Bruce Moran (ed.), *Patronage and Institutions: Science, Technology, and Medicine at the European Court 1500–1700* (Rochester: Boydell, 1991); Mario Biagioli, *Galileo, Courtier: The Practice of Science in the Culture of Absolutism* (Chicago: University of Chicago Press, 1993); Lisa T. Sarasohn, "Nicolas-Claude Fabri de Peiresc and the patronage of the new science in the 17th century," *Isis*, 84 (1993), pp. 70–90.

world and developed botanical gardens.[19] Even if these institutions are not the same as our institutions of "Big Science," they produced something that was radically different to what existed before 1500.

Universities have also attracted renewed attention. While they were not primarily responsible for the major innovations of the early modern period, they provided education and employment for mathematicians, natural philosophers, and physicians and institutional support in the form of botanical gardens, anatomical theaters, and cabinets of instruments.[20] Granted, it was not until the eighteenth century that universities really adopted and taught experimental physics and Newtonian mathematics and mechanics. Medical societies could also promote or hinder the spread of new approaches to natural philosophy and medicine, as Allen Debus shows in his essay.

As claims can only count as knowledge once they are transmitted and accepted, but are also transformed in the process of dissemination, the media of transmission are particularly worthy of study. Scholarly publishing changed dramatically in the sixteenth and seventeenth centuries. Correspondence served a vital role in the dissemination of knowledge and in claims of priority for discoveries, and much of it was intended for actual publication. In the mid-seventeenth century, Marin Mersenne (1588–1648) in Paris and Athanasius Kircher (1602–80) in Rome both sat at the center of vast networks that stretched around the globe. Kircher published much of what he received in his numerous books; Mersenne forwarded what he received on to other savants around Europe. By the end of the seventeenth century, the function of claiming priority and dissemination was being replaced by journals, in particular the *Philosophical Transactions*, the *Journal des sçavans* (both founded 1665), and the *Acta Eruditorum* (published in Leipzig from 1682). The two media flowed into each other; Henry Oldenburg (ca. 1619–77), the secretary of the Royal Society, himself sat at the center of a network of correspondence which supplied much of the *Philosophical Transactions'* material. All aspects of print culture, particularly the role of illustration, are also receiving greater attention.[21]

Social history focuses on everyday practice. Most modern scientists are not Nobel Prize winners, but patient contributors to a larger enterprise. Early

19   Steven J. Harris Harris, "Long-distance corporations, big sciences, and the geography of knowledge," *Configurations*, 6 (1998), pp. 269–304; Richard H. Grove, *Green Imperialism: Colonial Expansion, Tropical Island Edens and Origins of Environmentalism* (Cambridge: Cambridge University Press, 1995).

20   John Gascoigne, "A reappraisal of the role of the universities in the Scientific Revolution," in *Reappraisals of the Scientific Revolution*, eds David C. Lindberg and Robert S. Westman (Cambridge: Cambridge University Press, 1990), pp. 207–60.

21   Brian J. Ford, *Images of Science: A History of Scientific Illustration* (Oxford: Oxford University Press, 1993); William B. Ashworth, Jr, "Iconography of a New Physics," *History and Technology*, 4 (1987), pp. 267–97; Adrian Johns, *The Nature of the Book: Print and Knowledge in the Making* (Chicago: University of Chicago Press, 1998).

modern practitioners were no different. Yet if our account is to have any claims on completeness, their stories must be included. Professors of natural philosophy at the many minor universities, gentlemen who sent contributions to the scientific societies, noblewomen who produced and consumed literature popularizing and disseminating natural knowledge, and ships' captains who tested and improved methods of cartography and navigation all have a place in this more inclusive history of early modern science. Margaret Jacob tells us how the dissemination of the Scientific Revolution among broad classes of people in eighteenth-century England produced the Industrial Revolution.

This concern with practice has meant that the role of practical activities is being reassessed outside of a Marxist, externalist framework in order to break down simplistic dichotomies between pure science and artisanal or technological work.[22] Similarly, extending the history of science beyond conceptual developments to the practices of science has led to closer study of experimental natural philosophy. For the first time natural philosophers were constructing instruments that artificially created phenomena that did not exist or could not be observed in nature. The air-pump is the classic example, but experiment was not limited to physics; physicians carried out physiological experiments on animals, often to show that organs or muscles worked in mechanical ways. The development of experimental philosophy tied in neatly with the spread of the mechanical philosophy since most experimenters sought to provide mechanical accounts of the phenomena they investigated, such as Robert Boyle's suggestion that spring-like particles accounted for the elasticity of air.[23]

One of the main assumptions of newer scholarship is that current hierarchies of knowledge are not necessarily accurate when applied to other times. Rather, one should treat forms of knowledge symmetrically and without projecting onto the past the disciplinary boundaries of the present. The result is that disciplines which were once dismissed as pseudo-sciences have received fundamental reevaluation. The involvement of leading figures in these pursuits was once seen as an aberration to be explained away or simply ignored. But early modern practitioners' engagement with these disciplines was real and sincere. Newton's exhaustive engagement with alchemy is now seen as a fundamental part of his investigative program.[24] Concern with

22   Lisa Jardin, *Ingenious pursuits: building the scientific revolution* (New York: Nan A. Talese, 1999); J. A. Bennett, "The challenge of practical mathematics," in eds Stephen Pumfrey et al., *Science, Culture and Popular Belief* (Manchester: Manchester University Press, 1991).
23   Steven Shapin and Simon Schaffer, *Leviathan and the Air-Pump: Hobbes, Boyle, and the Experimental Life* (Princeton; Princeton University Press, 1985); Peter Dear, *Discipline and Experience: The Mathematical Way in the Scientific Revolution* (Chicago: University of Chicago Press, 1995).
24   Betty Jo Teeter Dobbs, *The Foundations of Newton's Alchemy: or, "The Hunting of the Greene Lyon"* (Cambridge: Cambridge University Press, 1975); and idem, *The Janus Faces of Genius: the Role of Alchemy in Newton's Thought* (Cambridge: Cambridge University Press, 1991).

improving the accuracy of astrological predictions was one of the major motivations behind the desire to improve astronomical models and observations. Copernicus, we now know, was motivated by a desire to respond to criticisms of astrology.[25]

In fact it is difficult to distinguish between the desire of natural magicians, such as Giambattista della Porta (1535–1615) and Heinrich Cornelius Agrippa von Nettesheim (1486–1535), to use their knowledge of occult sympathies to manipulate nature to useful ends and the Baconian program of using natural philosophy to benefit humanity. One could argue that science appropriated natural magic's goals while abandoning its similitudes and sympathies in favor of mechanical causation.[26]

Similarly the distinctions between a supposedly pseudo-scientific alchemy and a rational, scientific chemistry have become blurred. Simply, dichotomies between alchemy and chemistry, whether based on theory, method, purpose, or chronology, no longer hold. William Newman and Lawrence Principe have proposed using the term "chymistry" for both alchemy and chemistry in the early modern period to avoid imposing an arbitrary and anachronistic distinction on them.[27] There were alchemists of all stripes. They were prominent members of the Royal Society. There were Paracelsian alchemists who sought to reform medicine (see Debus's piece). There were alchemists who employed various forms of mechanical philosophy such as atomism. In fact, alchemy was probably at its height in the seventeenth century and it did much more than simply try to turn lead into gold.

Cultural and social approaches to early modern science have devoted considerable, if still inadequate, attention to issues of gender. Since all of the great figures of the Scientific Revolution were male, women were almost completely absent from the historical narrative. While nobody is suggesting that a long-lost female Newton will be rediscovered, women's activities in early modern natural philosophy are being recovered from historical silence.[28] Aristocratic women were particularly prominent as patrons and correspondents. Princess Elizabeth of Bohemia (1618–80) and Queen Christina of Sweden (1626–89)

25   See Robert S. Westman, *The Copernican Question* (Chicago: University of Chicago Press, forthcoming).
26   For an overview see John Henry, "Magic and science in the sixteenth and seventeenth centuries," in *Companion to the History of Modern Science*, eds R. C. Olby et al. (London: Routledge, 1990), pp. 583–96.
27   William R. Newman and Lawrence M. Principe, "Alchemy vs. chemistry: The etymological origins of a historiographic mistake," *Early Science and Medicine*, 3 (1998), pp. 32–65. See also Pamela H. Smith, *The Business of Alchemy: Science and Culture in the Holy Roman Empire* (Princeton: Princeton University Press, 1994).
28   Londa Schiebinger, *The Mind Has No Sex?: Women in the Origins of Modern Science* (Cambridge, Mass.: Harvard University Press, 1989); idem, *Nature's Body: Gender in the Making of Modern Science* (Boston: Beacon Press, 1993).

supported Descartes. Sophia, Electress of Hanover, (1630–1714) and her daughter Sophia Charlotte (1668–1705), who became Queen of Prussia, both conducted correspondence and conversation with Leibniz.

Despite the constraints on women, such as limited access to education and stereotypes accepted by both men and women about women's limited and flawed intellectual capacities, women were themselves active in the production of knowledge. Women worked in private astronomical observatories, for example. The astronomer Johannes Hevelius's second wife Katherina Elizabetha Koopman (1647–93) assisted with his observations and published two of his works after his death. Maria Sibylla Merian (1647–1717) traveled to the New World to collect specimens and write and illustrate works of New World natural history. Both these women, it should be noted, received their training in the private realm of the family. Gradually, however, as distinctions between the public and private sphere solidified, women's place was confined to the private, excluding them from participation in the increasingly public world of science.

More controversial is work on the gendering of nature and natural knowledge itself. Since cultural approaches to scientific knowledge argue that the scientific enterprise reflects the cultural values and commitments of practitioners, conceptions of gender should be reflected in scientific knowledge. It has been argued that the entire mechanical worldview is an inherently male one, repressing and replacing a more "female" organic view of nature.[29]

The history of early modern science has also gone beyond tired old stereotypes about the supposed conflict of science and religion to examine their interaction. In fact, a sensitivity to religion is a defining feature of much recent work on early modern science.[30] The Galileo affair continues to receive considerable attention, although the story is no longer seen by historians of science as an example of Catholicism's supposedly inherent hostility to science (see Westman's article).[31] As Merton suggested, religion played an important role in promoting the investigation of nature, but it is now clear it was not just Puritanism which did so.

In fact the study of God's creation, his Book of Nature, was widely regarded by philosophers and mathematicians of all confessions as a valuable way of overcoming skepticism and the atheism to which many thought it inevitably led. Although the argument from design did not receive that name until the

29   Carolyn Merchant, *The Death of Nature: Women, Ecology, and the Scientific Revolution* (San Francisco: Harper & Row, 1980).

30   For an overview see John Hedley Brooke, *Religion and Science* (Cambridge: Cambridge University Press, 1991).

31   The best overview of the Galileo affair is without doubt Annibale Fantoli, *Galileo: For Copernicanism and the Church* (Vatican City: Vatican Observatory Publications, 1996).

eighteenth century, it was already widespread in the seventeenth as a weapon against atheism. It argued that the order and purpose we see in nature could only be the result of a divine intelligence.[32] The Protestant Kepler, Jesuit mathematics professors, the founders of the Royal Society – all saw themselves as serving God by studying his creation.

As a result of these studies, the Scientific Revolution as concept and narrative is in turmoil. In fact, all three elements of the phrase are questionable. "The" implies that there was just one revolution in the early modern period, which was not the case. Moreover, it also implies a hierarchy of significance; yet if the events of the sixteenth and seventeenth centuries were "the" Scientific Revolution, then the chemical revolution of the eighteenth century was merely "a" revolution and necessarily of secondary importance, something that historians of other eras would dispute.

Furthermore, with greater sensitivity to actors' categories, historians now accept that the word "scientific" is also misleading and anachronistic. The association between science and the study of nature was not one made in the early modern period. In a narrow sense science, from the Latin *scientia*, meant any kind of certain knowledge, that is, it referred to the epistemological status of knowledge rather than its object. In a broader sense, *scientia* referred to any intellectual discipline rigorously practiced (usually at a university), including theology and logic. The various disciplines that studied and manipulated nature in the early modern period had many different terms: natural philosophy, mathematics, mixed mathematics, natural history, to name the most common. No single term encompassed these different enterprises in the way that "science" encompasses the study of nature today.

And "revolution" seems somehow inappropriate for a process that lasted at least a century and a half or, by many accounts, even longer. Furthermore, the lens of revolution is distorting; whatever doesn't measure up as a radical advance from our standpoint is tossed out of the historical story – something quite at odds with the current emphasis on respecting actors' categories. Thus the term "Scientific Revolution" has fallen out of favor without being replaced. "Early modern science," which brings less intellectual baggage with it, has had some success.

So what, then, should we do with the Scientific Revolution? Cunningham and Williams propose the radical solution of simply throwing the whole thing out. While they raise important arguments, their call has not found universal favor. The Scientific Revolution does retain considerable usefulness as a shorthand name for the field and it certainly shows no signs of simply disappearing.[33] This volume provides examples of a more moderate approach that

32   Amos Funkenstein, *Theology and the Scientific Imagination from the Middle Ages to the Seventeenth Century* (Princeton: Princeton University Press, 1986).
33   Witness the use of the term in several recent undergraduate survey texts: Steven Shapin, *The Scientific Revolution* (Chicago: University of Chicago Press, 1996); John Henry, *The Scientific*

retains the term as one describing the totality of developments in the period. But rather than attempting to define the exact essence of the Scientific Revolution, they adopt one or more of the approaches outlined above: locating practitioners in their cultural context; revealing how knowledge is socially created and transmitted; expanding the narrative to include more disciplines; examining the connections between science, craft, art, and technology. They show us that whether we want to use the term revolution, whether we accept that early modern science was not the same as ours, the study of nature was fundamentally transformed between 1500 and 1700, not just in its theory, but in its methods, institutions, and everyday practices.

Revolution and the Origins of Modern Science (New York: St Martin's Press, 1997). Peter Dear, *Revolutionizing the Sciences: European Knowledge and Its Ambitions* (Princeton: Princeton University Press, 2001), reverses and pluralizes the terms.

1

A Traditional Narrative of the Scientific Revolution

The Rise of Modern Science: When and Why?

R. Hooykaas

Originally appeared as "The Rise of Modern Science: When and Why?" (*BJHS* 1987: 453–73).

Editor's Introduction

Reijer Hooykaas belongs to the generation that introduced the history of science to universities in the United States and Europe. His essay provides us with an excellent starting point for our readings on the Scientific Revolution. For Hooykaas there is no doubt that the sixteenth and seventeenth centuries witnessed a dramatic improvement in what people knew about nature and in the way they studied it. He grants that all discussion about the origins or rise of science depend heavily on the definition of science used, but is confident that the origins of modern science are to be found in the early modern period. Here he is arguing against the claim first put by the French physicist and historian Pierre Duhem that the origins of modern science can be found as early as the thirteenth century.

In Hooykaas's view two developments were most responsible for the rise of science in early modern Europe. The first was the process of discovery begun by the voyages of Portuguese navigators that culminated in the discovery of the New World. The existence of lands, plants, animals, and people in Asia, Africa, and the New World that were not mentioned in any ancient sources undermined their authority. Experience, Hooykaas argues, now took precedence over human authority and reason alone. The second major development was the transition from an organic view of the world to a mechanical one. Nature worked like a machine, consequently it could be studied like one, that is, mathematically.

In Hooykaas's essay we can see a reaction against a view of science as primarily a cognitive activity, one that was essentially distinct from society and other cultural activities. This view held that the task of the historian of science was to trace the red thread of truth through the centuries, while identifying and then pruning away the false shoots of erroneous theories. Hooykaas terms this approach the teleological or progressionist method. Such historians also argued that a dramatic break occurred in science in the sixteenth and seventeenth centuries, but it was a development that was internal to science and not the result of broader historical events. In contrast Hooykaas insists that in order to understand the rise of science we also need to examine factors which create a favorable climate for science, ranging from theological concerns to the rise of the artisanal classes.

Although most current historians of science share Hooykaas's rejection of the teleological approach, many go far beyond him. For example, although Hooykaas emphasizes the importance of facts in the rise of modern science, he does not explore what facts actually are or where they come from. Our later pieces explicitly take up these questions, arguing that the production of scientific facts is in itself a social process. Other essays go beyond Hooykaas's emphasis on the mechanical and mathematical disciplines to include natural history, medicine, and alchemy.

The Rise of Modern Science: When and Why?

R. Hooykaas*

When did modern science arise?[1] This is a question which has received divergent answers. Some would say that it started in the High Middle Ages (1277), or that it began with the 'via moderna' of the fourteenth century. More widespread is the idea that the Italian Renaissance was also the re-birth of the sciences. In general, Copernicus is then singled out as the great revolutionary, and the 'scientific revolution' is said to have taken place during the period from Copernicus to Newton. Others would hold that the scientific revolution started in the seventeenth century and that it covered the period from Galileo to Newton. Sometimes a second scientific revolution is said to have occurred in the first quarter of the twentieth century (Planck, Einstein, Bohr, Heisenberg, etc.), a revolution which should be considered as great as the first one.

It might be asked: Was there ever something like a scientific *revolution*? Perhaps the term is not well chosen; with the word 'revolution' we usually connect the idea of 'revolt', 'violence' or, at any rate, abruptness; a 'revolution' covering 100, or 150, or 200 years hardly deserves that name.

It must be recognized, of course, that there is an enormous gap between the science of Antiquity and the Middle Ages on the one hand and that of the seventeenth century onwards on the other. Even without analysing their respective contents, their *effects* convincingly show the watershed: on the basis of 'ancient' science one cannot construct locomotives or aeroplanes; on the basis of 'modern' science this has turned out to be possible. The gap between ancient

1 The present author has dealt with some aspects of the problem in earlier publications: (a) 'Science and theology in the Middle Ages', *Free Univ. Qu.* (1954), 3, pp. 77–163 (in particular, pp. 77–82, 88–97, 103–118, 131–137); (b) *Religion and the Rise of Modern Science*, Edinburgh and London, 1972 ff (pp. 9–13, 29–41, 61–66, 88–94); (c) *Das Verhältnis von Physik und Mechanik in historischer Hinsicht*, Wiesbaden, 1963 (reprinted in: R. Hooykaas, *Selected Studies in History of Science*, Coimbra, 1983, pp. 167–199); (d) 'Von der "physica" zur Physik', in: *Humanismus und Naturforschung*, Beitr. Humanismusforschung VI, Boppard, 1980, pp. 9–38 (reprinted in: *Selected Studies*, op. cit., pp. 599–634); (e) *Science in Manueline Style, the Historical Context of D. João de Castro's Works*, Coimbra, 1980 [pre-print of: *Obras Completas de D. João de Castro* (ed. A. Cortesão e L. de Albuquerque), vol. IV, Coimbra, 1982, pp. 231–426 (in particular pp. 3–12, 92–98, 146–152, 163–167)].

* Originally intended to be the closing lecture of the Fourth Reunion of the History of Nautical Sciences, Sagres Portugal), 3–8 July 1983, under the presidency of Professor L. de Albuquerque (Coimbra and Lisbon).

science and Newtonian science is wider than that between Newtonian classical – modern science and Planck – Bohr physics. In the latter case there is a large measure of continuity with the preceding epoch: the new physics does not render the classical – modern invalid, whereas Newtonian and scholastic physics are quite incompatible.

It should be emphasized that before expounding our opinion about the question 'When and why did the great scientific change take place', we have to realize that the answer depends on the (often hidden) methodological principles applied by the historiographer of science, and – secondly – on which characteristics are deemed to constitute 'modern science'.

Historiographical Methods[2]

The approach most attractive to modern scientists is one that could be called 'evolutionistic, 'teleological', 'genealogical' or 'progressionist'. Without any doubt, science bears a cumulative and progressive character, and this confronts a historian of science with problems the historian of art or of theology hardly meets with. Consequently, the historian of science tends to apply to the past the standard of present-day knowledge. Starting from the present-day advanced state of science, the evolutionistic historian goes back into the past with all its conflicting and erroneous opinions, its fertile and its sterile theories, its thoroughfares and its blind alleys, taking the red thread with him, until he arrives at the 'fathers'. Next – and then begins his serious history-writing – he follows backwards his thread of progress, discarding all that does not directly lead to the exit of the maze as aberrations and errors, until he returns to the present situation. He concentrates on the heroes of science and has a keen eye for the 'precursors' who paved the way for them. The past serves mainly to prepare and announce the present: almost unwittingly, a teleological conception dictates the shape of his historiography.

It is understandable that the scientist who writes about the history of his discipline has a predilection for the path that led to present-day science, for he is more interested in finding out how his predecessors escaped from the maze than in studying their unsuccessful efforts.

---

2 On this topic, see: (a) E.J. Dijksterhuis, *Doel en Methode van de Geschiedenis der Exacte Wetenschappen*, Inaugural address, Utrecht, 1952, pp. 13–20; (b) R. Hooykaas, *L'histoire des Sciences, ses Problèmes, sa Méthode, son But*, Coimbra, 1962 (also in: *Rev. Fac. De Ciências Coimbra* (1963), 32, pp. 5–35); (c) T. Frangsmyr, 'Science or history, G. Sarton and the positivist tradition in the history of science', *Lychnos* (1973–1974), pp. 104–144; (d) F. Krafft, 'Die Naturwissenschaften und ihre Geschichte, *Sudhoff's Arch.* (1976), 60, pp. 317–337; (e) R. Hooykaas, 'Pitfalls in the historiography of geological science', *Nature et Histoire*, (1982), 19–20, pp. 2–33; (f) R. Hooykaas, 'Wissenschaftsgeschichte, eine Brücke zwischen Natur und Geisteswissenschaften', *Berichte z. Wissensch. Geschichte* (1982), 5, pp. 162–170.

In sharp contrast to the progressionist approach is a more phenomenological and imaginative one. It considers the historian's task to be to revive the past, to enter into the minds of our predecessors, to imagine the political, social and cultural aspects of their environment, to re-enact their metaphysical, ethical and scientific conceptions and to identify himself as much as possible with their personalities. Standing beside them in the centre of the maze, he then enters with them into the blind alleys, too; he positively appreciates their accounts of the then known facts, fitting them into the then generally accepted theoretical system, even when afterwards these theories turned out to be false. He will then recognize that the way out of the labyrinth could hardly have been found without coming to many dead ends. He will also recognize that by entering the blind alleys, our predecessors erected for us warning-posts reading 'no thoroughfare'. This re-enactment of the past will show him that, however incomplete the knowledge of facts and however obsolete the opinions held, the *method* of these pioneers was often 'scientific', even when the result was not 'true'. Scientific results of the past may *now* be known not to conform to physical reality, but these same results may have been conformable to what was *then* considered to be physical reality, as they were obtained by keen observation and consistent thinking.

The distinction between the two approaches should not be considered absolute: because of the cumulative character of science, the historian who is also a scientist cannot leave out of sight the genealogy of present-day theories and concepts. He goes wrong, however, if he thinks that not only scientific knowledge but also the quality of scientific *thinking* has improved and that our predecessors were more primitive or less clever than we are.

The Character of Modern Science

When speaking about the rise of modern science it is necessary first to state what we take to be its characteristics.

1   Modern science acknowledges no authorities (however great they may be) except the authority of nature itself. It does not even acknowledge the authority of the investigator's own reason. In case of a conflict between his rational expectations and his discoveries by observation, the investigator's reason must adapt itself to the data provided by nature. As T. H. Huxley put it: 'Sit down before fact as a little child . . . follow humbly wherever and to whatever abysses nature leads, or you shall learn nothing'. In modern science a rational and critical empiricism triumphs over rationalism (self-sufficiency of theoretical reason).

2   Modern science is experimental. It is built not only upon direct observation of nature, but also upon artificial experiments, conquering

nature by art and obtaining genuine information from it through interference by art.

3   Modern science favours a mechanistic world picture, explaining natural
phenomena as much as possible by analogy with a mechanism. Ancient
science, on the other hand, tended to an 'organistic' world view, regarding non-living things as to a large extent similar to organic beings.

4   Modern science tries as much as possible to describe or explain natural
things and events in mathematical terms and to quantify qualities.

It should be stressed that these characteristics are not wholly absent from
ancient science. It had its observations, experiments, mechanistic interpretations and mathematical descriptions, but they did not play so predominant a
role as in modern science, although, of course, in the 'mathematical arts'
(which, however, did not belong to the 'philosophy of nature' or 'physics')
mathematics played an important role.

In order, therefore, to locate the transition from ancient to modern science
we have to concentrate our attention on these characteristics. This is no easy
task, for some disciplines may show only one or two such characteristics and
yet give the impression of 'modernity'. Consequently, there is some vagueness
in the data, and difference of evaluation is inevitable. At any rate, in this
essay we will not deny the name of 'modern science' to disciplines that show
little mathematization (as was until quite recently the case with geological
and biological sciences), or which occupy themselves more with classification
than with causal explanations and measurement of quantities. When tack-
ling the problem of the 'scientific revolution' we must consider the whole range
of sciences of nature and not only the mathematical–physical disciplines
(astronomy, physics).

In our search, the following influences might be taken into acount: (1)
empiricism; an emphasis on empirical reality over and above speculative reasoning; (2) analysis of phenomena in an experimental way, as against a purely
logical analysis; (3) establishment of a science free from the constraint of
authority, except that of nature herself; and (4) mathematization of nature
and measurement of phenomena.

Moreover, any historical events creating a favourable climate for science
should be taken into account, such as: (a) the emergence of theological voluntarism, in opposition to intellectualism; (b) the emergence of mechanistic
conceptions over against organistic ones; (c) the emancipation of manual
workers and acceptance of manual experiments; and (d) the extension of
natural history on the basis of experience rather than book-learning, triggered
off by the 'geographical revolution'.

Some of these factors had no immediate and direct effect in science, but
they created an atmosphere favourable to the reception of new ideas and
methods.

The Middle Ages

All these preliminary remarks warn us that it will be difficult to find a hard and fast answer to the question 'Why and when did modern science begin?' Nobody can completely free himself from subjective predispositions, particularly with regard to general problems like the present one. Nevertheless, we can try to be as objective as possible.

The French physicist and historian of science, Pierre Duhem, has often been accused of giving in to nationalistic prejudices. He claimed that modern science was born in 1277, when the bishop of Paris, at the instigation of Pope John XXI, condemned a great many theses that introduced Greek necessitarianism into theology, putting the Necessity of Nature above the sovereign will of God:

> Étienne Tempier et son conseil, en frappant ces propositions d'anathème, déclaraient que . . . pour ne pas imposer d'entraves à la toute-pouissance de Dieu, il fallait rejeter la Physique péripatéticienne. Par là, ils réclamaient implicitement la création d'une Physique nouvelle . . . Cette Physique nouvelle, nous verrons que l'Université de Paris, au XIVe siècle, s'est efforcée de la construire et qu'en cette tentative, elle a posé les fondements de *la science moderne; celle-ci naquit, peut-on dire, le 7 mars 1277, du décret porté par Monseigneur Étienne, Évêque de Paris*; l'un des principaux objets du présent ouvrage sera de justifier cette assertion'.[3]

Such a precise identification of the date of birth of a modern science is rare, although not unique.

It must be recognized that there is *some* truth in Duhem's verdict. Of course, the bishop was not acting here as an advocate a new science, but as a defender of the ancient biblical faith. He set Christian voluntarism over and above philosophical intellectualism. The then new Aristotelian philosophy (put forward in its most radical version by the Averroïsts) decreed a priori on allegedly rational grounds, that if the heavens were to cease turning round, no change on earth would be possible ('tow would not burn'), that God cannot create new species, and that only uniform, circular motions are possible in the heavens, etc. All these prohibitions or limitations to the natural world were proclaimed because these things were deemed to be intrinsically against the eternal order of Nature and against Reason. Tempier, however, maintained that God's will is more powerful than Nature or pretendedly 'eternal Reason'. He did not say that new species do arise, or that rectilinear motions do occur in the heavens, but only that human reason has no right to put any limits to God's power. This implies that natural science cannot decide a priori with absolute certainty how nature ought to be and must be found to be, but that

3   P. Duhem, *Le Système du Monde, VI*, 2nd edn, Paris, 1954, p. 66.

we have just to accept phenomena as it has pleased the Creator to give them, whether they seem conformable to human reason or not. This is a very important metaphysical standpoint (emphasizing the contingency of nature), which certainly is not anti-scientific. One wonders, therefore, how the historian Lynn Thorndike could dub it 'warfare (of theology) with science' and compare it with the 'silencing' of Galileo.[4]

In the Paris decree, a purely theological issue was at stake, but an issue that could indeed have great, positive consequences for the freedom of scientific theorizing and for the choice between a rationalistic and an empiricalist approach to nature. For the time being, however, it had no influence on scientific speculation, let alone on practical research. Although the Aristotelian world picture was deprived of its absolute authority, it was not seriously criticized or replaced by an alternative system. Tempier's decree could have created a favourable spiritual climate for a more empirical science, but it had no direct consequences in that field. Therefore, it could hardly be called the '*birth* of modern science'.

Of course, Duhem knew that quite well, for he said also that the consequences were not realized until the fourteenth century (i.e. 100 years later) at the university of Paris. He alluded here to nominalism, the 'via moderna', which by its critique of Aristotelian philosophy had an enormous influence. The nominalists emphasized the contingency (the not rationally deducible, not necessary, character, the just-given-ness) of the world, which has been made by God's incomprehensible will; only a posteriori can we put together a science of nature, as rational as possible in our own eyes; and such a system will be at best highly probable though not absolutely necessary, since God could have willed a different world with different rules.

Nominalism was against eternal Forms or eternal species; a 'universal' is but a name for a group of individuals we deem to belong together. The thoroughgoing philosophical empiricism of the nominalists was thus based on theological voluntarism. Small wonder, then, that the main Parisian protagonists of the 'via moderna', Jean Buridan (*c.* 1300–1358) and Nicole Oresme (*c.* 1323–1382), more than once referred to the decree of 1277 in order to back up their standpoint.[5]

Because of the nominalists' introduction of the impetus theory, their mathematization of physical problems, and their undermining of the dividing wall between terrestrial and celestial mechanics, Duhem on one occasion becomes so enthusiastic that he seems to forget that he has dated the beginning of modern science with Tempier's decree of 1277, and gives the honour to Buridan. He now says that a sharp line separates ancient science from modern

4   L. Thorndike, *A History of Magic and Experimental Science, III*. New York, 1934, p. 470.
5   J. Buridanus, *De Coel et Mundo*, lib. I, qu. 16, N. Oresme, *Le livre du Ciel et du Monde (c.* 1370), l. II, 95c (ed. A.D. Menut and A.J. Denomy), Madison, 1965, p. 374.

science, viz. the moment when the stars were no longer held to be moved by divine beings, but heavenly and sublunar motions were considered to depend on the same mechanics:

> si l'on voulait, *par une ligne précise*, séparer le règne de la Science antique du règne de la Science moderne, il la raudrait tracer, croyons-nous, à l'instant où Jean Buridan a conçu cette théorie [de l'impetus], à l'instant où l'on a cessé de regarder les astres comme mus par des êtres divins, où l'on a admis que les mouvements célestes et les mouvements sublunaires dépendaient d'une même mécanique.[6]

According to Duhem, the impetus theory paved the way for Galileo's mechanics and thus marked the beginning of modern physico-mathematical science. Although Buridan may have made (all) change measurable in principle, however, we should note that he and his disciples did not perform any measurement.

It should also be remembered that mathematization and measurement are no prerogatives of modern science alone. Ancient astronomy combined a highly sophisticated mathematical description of heavenly motions with rather exact measurements. In statics and hydrostatics, too, measurements had long been performed. However, those were reckoned to belong to the 'most physical part of mathematics' rather than to physis in the proper sense; that is, they did not belong to 'science' (philosophy of the nature – physis – of things) but rather to the 'lower' of the liberal arts. Nevertheless, today we recognize them as an important part of physical science. In modern times these mathematical 'arts' (astronomy, mechanics, etc.) have become part of science, whereas speculative natural philosophy has been more and more relegated to the periphery. The acceptance of these arts as more than a mere auxiliary part of physical science, however, was to be a difficult process.

Much later, the alliance between theoretical reason and manual experimentation was to play a very important role in the rise of modem science. The decree of 1277 and fourteenth-century nominalism, however, did not at the time cause the slightest change in the physical methods of those who understood their message. The changes caused by the 'calculatores' and the Parisian school needed something from the outside to make them bear fruit.

Natural History

When the Portuguese seafarers discovered that the tropical regions were habitable and inhabited, that there was much land south of the equator, that

---

6. P. Duhem, *Études sur Léonard de Vinci*, III, p. XI; II, p. 411. Cf. E. Gilson, *La Philosophie au Moyen Age*, 3rd edn, Paris, 1947, pp. 460, 487.

there was more dry land on the globe than had been taught them, that Southern India protruded much farther into the 'Indian Sea' than Ptolemy had told them and that the shape of West Africa (the Gulf of Guinea) was widely different from what ancient maps indicated – all this gave a severe shock not only to them but to the learned world as well. Ptolemy, the great authority in astronomy throughout the later Middle Ages and (since the recent discovery of his 'Geographia') the greatest authority in geography, too, now turned out to be not wholly reliable. He might be a great mathematician, but his 'natural history' was not so good. The same was the case with all those writers of Antiquity who had described peoples, animals and plants. There were many things whose existence they had not known and also many things they had 'known' wrongly. Their knowledge was incomplete and often erroneous.

With these simple seafarers a new natural history arose. They discovered, as Pedro Nunes (1537), following Policiano (1491) put it: 'new islands, new countries, new seas, new peoples and what's more, new heavens and new stars'. Just at the time when humanism (which wanted to go *back* from the 'barbarous' and 'gothic' period to the perfection of the Ancients) was penetrating into Portugal, their own *experience* taught the sailors that those glorified and quasi-infallible Ancients were as fallible and as human as their contemporaries.

The early Portuguese navigators (Diogo Gomes, 1460; Duarte Pacheco Pereira, 1506) testified to how amazed they were by the things seen during their voyages, and time and again they protested that, however strange these phenomena might seem to be, they had 'seen them with their eyes and touched them with their hands'. There are echoes here of the apostle Thomas, whose reason refused to accept the resurrection of the Lord until, as St John relates, having seen and touched Him, he was convinced: experience had overcome aprioristic reasoning.

In the competition between Reason and Experience, the precedence was now reversed. The navigator D. João de Castro (1500–1548) wrote that whereas formerly the existence of antipodes was deemed to be against reason, now that the experience of Portuguese seafarers had proved their existence, it had become 'a thing most conformable to reason'.[7] That is: we do not put experience to the test of theoretical reason, but we submit theoretical reason to the test of experience. These pioneers, who were not hindered by learned prejudices, did not make their decisions whether a certain fact was true by arguments *pro et contra*: for them observation was enough, and facts must be accepted in spite of any apparent 'absurdity'.

Both the scholastic philosophers and the humanists, who tenaciously clung to ancient traditions, were deeply shocked, and at first they tried to

---

7   D. João de Castro, *Tratado da Esfera* (*c.* 1538), Obras I, p. 58.

save the honour and authority of the Ancients by various exegetical tricks. But it was all in vain, for the evidence of the facts was too strong. Most irksome for them was that all this new and subversive information was adduced by unlearned sailors. These they held in low esteem; yet now it was precisely these uneducated people who had put them to shame. On the other hand, some of them took the lesson to heart. Peter Ramus (1546) wrote: 'The philosophers, orators, poets and scholars of the whole world and of so many ages did not know what navigators, merchants, uneducated people learned, not by arguments but through experience ... we are compelled by simple examples and immediate experience of the senses to recognize that those very ancient prodigies of wisdom have at last lost their monopoly and have been outdone'.[8] And indeed, as Camões ironically pointed out, there were things he had seen which the 'uneducated sailors' who had only experience as their teacher had proved to be true, yet which those who investigated the secrets of the world by their sharp wits alone had demonstrated to be wrong.

This marks the beginning of *a new, empiricist, non-rationalistic trend in science*: problems are solved by reasoned experience and not by scholastic discussions, which – however clever and logical they might be – brought forth only armchair physics.

The Emancipation of the Burgher Class

All this happened at a time that could hardly have been more favourable. It coincided with the emancipation of the burgher class, in particular in Italy, Southern Germany and the Netherlands. The artisans of the late Middle Ages and early Renaissance became conscious of their dignity and social importance. This self-respect, also in intellectual matters, was evident in people like the Huguenot potter Bernard Palissy (1510–1590), who proudly declared that, although he had not read the books of the great Greeks and Romans and spoke only his mother tongue, he nevertheless had a right to contradict their reported opinions: 'I have read no other book but heaven and earth, which is known to everybody, and it has been given to everyman to know and to read this beautiful book'. Artisans like Palissy, Robert Norman and Albrecht Dürer wrote books – often more lively than the stylish works of the professional scholars – in which they related the results of their personal investigations and the interpretations they gave to these.

Like the great philosophers of Antiquity (Plato, Aristotle, Cicero, Seneca), many of their humanist followers looked down upon the mechanical arts and

8   P. Ramus, 'Scipionis Somnium' (1546), in: *Petri Rami Praelectiones in Ciceronis Orationes octo consiliares*, Basileae, 1580, p. 53.

those who cultivated them, the 'mechanical' workers, engineers, chemists, metallurgists, sailors, etc. The liberal arts, which did not require manual labour, were the only ones befitting a free citizen and a philosopher.

It seems, however, that among scholars in the fifteenth century, respect for the trades was growing in some parts of Europe. Nicolaus Cusanus (the son of a fisherman) in his 'De staticis experimentis' allows a scholar to be instructed by an un-lettered man (the 'Idiota'), a mechanician who tells him how some difficult practical problems may be solved by 'mechanical' means. Luis Vives (1492–1540), when living in the Southern Netherlands, advised students to follow the example of the fifteenth-century Louvain scholar Carolus Virulus, who sought contact with the fathers of his students in order to learn from them about their trades of cobbler, skipper, etc. He deplored that there was no 'history of the arts', the writing of which would be 'an occupation worthy of a burgher'. Although himself an accomplished humanist, Vives shows the burgher influence of his Flemish dwelling-place. He recognized that peasants and craftsmen were often closer to reality than his fellow-scholars, and that they knew nature better than those 'great philosophers' who, in place of real things (about which they were ignorant) imagined another nature, consisting of 'Forms, Ideas, and other chimerae'. In the wake of Vives, Peter Ramus (1515–1572) sought contact with artisans (instrument-makers, painters, etc.) and frequented their workshops in search of information about applications of mathematics.

Many sixteenth-century scholars were at the same time artisans (printers, engravers, instrument-makers). The cartographer Gerard Mercator engraved maps with his own hands; the Nuremberg clergyman Georg Hartmann (1489–1564) was not only an able mathematician and an experimenter on magnetism but also a good mechanician who made astrolabes, globes and sundials.

Mechanicism

The engineers of Antiquity (e.g. Hero of Alexandria) often used 'mechanistic', non-teleological, explanations of the phenomena they evoked artificially, and of 'natural' phenomena as well. The Renaissance period saw a slow penetration of their procedures and ideas into more philosophical and scholarly works, as a consequence of a closer contact between the two groups. 'Mechanical experiments and mechanistic interpretations (even of natural phenomena) became more common. Mechanicians always showed a tendency to make models of natural things and events (globes, planetaria, models of volcanoes). The thirteenth-century author of a book on the magnet, Petrus Peregrinus (1269), who in his experiments was 'ahead of his time', conceived of the outer heaven as a huge magnet, and speculated that an artificial spherical magnet that imitated the heavenly globe would by a sympathetic influence also turn round.

The social changes of the fifteenth and sixteenth centuries went together with a philosophical change: mechanical methods and models inevitably led to mechanistic explanations of phenomena. The organistic world view (which sought to understand all things by analogy with living beings) was penetrated and eventually replaced by a mechanistic world view, which tended to consider even living beings, as far as possible, as analogous to mechanisms. In general, this penetration had nothing abrupt and revolutionary about it. D. João de Castro (1538) applied 'modern' methods in his experiments on earth magnetism, but as he considered his measurements to belong to 'the lowest and most forgotten part of mathematics', this had no influence on his general world view, which was that of an old-fashioned Aristotelian, free of nominalistic taints.

In the early seventeenth century, the physician Angelo Sala (1617) synthesized copper vitriol. He interpreted his artificial product as a mechanical structure, an 'apposition of particles' of the ingredients he had used. This led him to the idea that *natural* vitriol, having the same properties as the artificial product, must also be 'an apposition of particles'. But sea-salt, which he could not synthesize, he considered as a 'unity', perfectly homogeneous, existing under its own specific Form. In the case of this chemist, then, the old philosophy was abandoned only in so far as the *facts* compelled him to do so.[9]

Such mechanistic explanations inevitably undermined the old, organistic, world view. If parts of nature are like mechanisms, it must be possible to *fabricate* them; if they were like organisms – which are propagated only by *generation* – they would be inimitable. So Sala thinks of vitriol in mechanistic terms, as something that can be fabricated, whereas sea-salt he holds to be generated by nature alone. For those natural events which can be artificially reproduced, knowledge of their 'nature' could now be obtained by experiment. We can speak here of an '*experimental philosophy*'.

From the end of the fifteenth century' (da Vinci) to the beginning of the seventeenth (Sala, Sennert, Basso), the motions, arrangements, sizes and shapes of invisible particles played an increasing role in scientific interpretations. About 1600 there was an outburst of these 'corpuscularian' theories, which ended in the comprehensive mechanistic systems of Gassendi, Descartes and others, systems from which the Forms and Ideas that so much annoyed Luís Vives had completely disappeared. It should be stressed, however, that the

9   Cf. (a) R. Hooykaas, *Het Begrip Element in zijn historisch-wijsgeerige ontwikkeling*, thesis: Utrecht, 1933, pp. 148–154; (b) 'The discrimination between natural and artificial substances and the development of corpuscular theory', *Arch. Intern. Hist. Sci.* (1948), 4, pp. 840–858; (c) 'The experimental origin of chemical atomic and molecular theory before Boyle', *Chymia* (1949), 2, pp. 65–80 (reprinted in: *Selected Studies*, op. cit. (1), pp. 285–308).

heuristic value of these corpuscular theories was not great. They 'explained' a posteriori, but they hardly predicted any phenomena. Nevertheless, they gave great support to the mechanistic picture and thus inspired confidence in 'mechanical' (experimental) research.

The work of mechanicians who co-operated with scholars (or were scholars themselves) led to the rise of what was called experimental philosophy (emphasizing the *method* used) or mechanical philosophy (referring to the scientific *models* used), and also to the development of a 'history of the arts' as part of the 'history of nature'. Empiricism (acceptance of facts), experimentalism (eliciting secrets from nature by mechanical means), and mechanicism (interpretation by means of models and images borrowed from mechanics rather than from living beings) all went together in the early seventeenth century. Only certain parts of the new mechanistic world picture, however, could be mathematized.

It is precisely on these mathematizable parts – the measurable, macroscopic phenomena of falling bodies, projected bodies and rotating bodies – that many historians concentrate their attention when considering what Anneliese Maier (1938) termed 'the mechanization of the world picture', which took place in the seventeenth century from Galileo to Newton. They consider this the most weighty factor in the rise of modern science.

Much can be said for this opinion, although there is the risk that the scientific 'revolution' is identified with the rise of modern mechanics, this discipline having become the heart of physics. The mathematization of kinematic and dynamic phenomena, which had been unsuccessful in the hands of the late medieval 'calculatores' (Suisseth) and protagonists of the 'latitude of forms' (Oresme), now at long last was linked with quantitative experiments. There is a clear progression from the speculations of these medieval scholars, via the sixteenth-century Italian engineers (Tartaglia, Benedetti) and Galileo, and then Descartes, Borelli and Huygens, finding its fulfilment with Newton. In this sequence a great step forward was made by Galileo at the beginning of the seventeenth century.

Merged in this sequence is the development of seventeenth-century astronomy under Kepler, Galileo and Huygens and, finally, Newton, who fitted the Copernican model of the universe definitively into a mechanistic system of nature. In his 'Principia' (1687), the synthesis of astronomy and physics, the mathematization that united terrestrial and celestial mechanics, was finally accomplished.

This story has the advantage of presenting a clearly continuous development and it rightly implies that the really 'modern' phase of physics started with the mathematical-descriptive work of Galileo and Kepler, and the outburst of explanatory corpuscular theories at the beginning of the seventeenth century. However, most people who accept the pattern depicted above, and hold that the 'scientific revolution' in mechanics took place about

1600, nevertheless see its astronomical root – and the real beginning of the revolution – in the publication of Copernicus' main work in 1543 – i.e. in an age of chaotic competition between various world views (Aristotelianism, Platonism, Pythagorism, Hermeticism). By the beginning of the seventeenth century, on the other hand, a certain unity had begun to emerge: both the mechanistic conception and the heliocentric theory were on their way to victory.

Copernicus was no adherent of a mechanistic world view, and his way of thinking was that of the Ancients rather than of the Moderns; yet he is inserted in the series of modern astronomers, because it was *his* model of the universe that formed the basis of the work of the great innovators, Kepler and Galileo. Understandably, this 'insertion' of Copernicus into the series has led to the widespread belief we have noted, that he was a 'revolutionary' who overturned the ancient dogmas and inaugurated 'modern science'.

Copernicus

Should Copernicus thus be regarded as the initiator of modern science, the first in the series of heroes of the scientific revolution: Copernicus – (Tycho) – Kepler – Galileo – Huygens – Newton? In order to answer this question it seems useful to consider first the 'novelty' of his work and its scientific character, and secondly its influence and evaluation during the sixteenth century and the evaluation by modern scholars.

Copernicus' dissatisfaction with the Ptolemaic system was hardly caused by its factual errors: large parts of Copernicus' data were borrowed from his ancient and medieval predecessors and he did not claim that his own measurements were more exact than theirs. But Ptolemy had introduced the equant (i.e., movements uniform with respect to a point away from the centre of the circular path). This was a deviation from the 'Platonic' rule that in astronomy the motions of the planets have to be reduced to combinations of perfectly *uniform* circular motions. To Copernicus, this violation of uniformity was the first reason why he became dissatisfied with the current system. The second was its lack of harmony: for each planet there had been introduced a specific set of circular motions, which had no interconnection with those of the other planets.

Now Copernicus, as a typical humanist, began perusing the works of the Ancients in order to see whether better solutions had been given in the past. He found then the daily motion of the earth mentioned by some Pythagoreans, whereas he seems to have believed that they also accepted the annual revolution of the earth round the sun (which at any rate he had found expounded by Aristarchus of Samos). He was thus able to appeal to the most ancient sources, and his contemporaries recognized this by dubbing his world system 'Pythagorean'.

As to the daily rotation of the earth, Copernicus knew quite well that 'mathematically' speaking it made no difference whether the heavens turn round in twenty-four hours or whether the earth does so. But physically it made a great difference, and as he claimed that his system was conformable to physical reality he had to offer a *physical* alternative to the Aristotelian arguments for the immobility of the earth. It was no new problem, for it had been a frequent topic in scholastic discussions. Most of Aristotle's and Ptolemy's arguments had already been answered in the fourteenth century by Nicole Oresme, but this philosopher finally rejected the earth's motion. The idea of the earth's rotation had the great disadvantage that it broke the unity of the scholastic philosophy of nature by maintaining that the natural motion of the four elements was not rectilinear but *circular*, and that falling heavy bodies moved towards the earth not because they wanted to approach the centre of the universe (as the Aristotelians held), but because a separated piece of 'earth' wanted to be united with the planetary globe to which it belonged. (In the same way, according to Copernicus, a piece of lunar matter would try to unite itself with the Moon.[10])

Copernicus' greatest achievement was his introduction to the annual motion of the earth. The apparently retrogressive motions of the planets could be attributed to this. Instead of a specific set of two circles (deferent and epicycle) for each planet, now one of these two was recognized to be the same for all planets: a projection of the earth's orbit. In this orbit Copernicus now found the 'common measure' of all other planetary motions (1514). Thus, a 'certain bond' of harmony in the universe, which he had so sorely missed in the 'vulgar' system, had been found. It should be mentioned, however, that some decades later Tycho Brahé (1588) reached the same result in his geo-heliocentric system, although without abandoning the immobility and central position of the earth.

Having expounded his cosmological system and physical tenets in the first Book of 'De Revolutionibus' (1543), Copernicus dealt with the astronomical calculations in the Books II–VI. Although he had explained the 'second inequality' of the planetary motions, he still had to account for the 'first inequality', while discarding the equants. In order to reach this aim he had again to resort to the traditional device of combining circular motions. The greater simplicity of his system suggested by the famous diagram in the first Book could not be maintained in the other five.

The equant had not been the only fly in the ointment of the current astronomy: several minor irregularities had crept into the system during the Middle

10   Cf. R. Hooykaas, *Science in Manueline Style*, op. cit. (1), pp. 50–61.

Ages. It was found, for instance, that the slow uniform motion of the equinoctial points along the ecliptic ('praecession') was subject to superimposed oscillatory movements ('trepidation') which, apart from disturbing the uniformity of the progression, were not circular. Copernicus now managed to reduce these motions (as well as some other irregularities in astronomy) to a combination of two uniform circular motions.

If we consider Copernicus' work from the standpoint of most sixteenth-century scholars, we would assume an amused or sceptical or even hostile attitude towards the contents of the first of the six books 'On the Revolution of the Heavenly Orbs', whereas we would praise the other five as an outstanding contribution to one of the important liberal arts. These books were highly appreciated as a 'mathematical' account of the heavenly motions, while strictly keeping to the ancient fundamental rule of astronomical art, viz. that only perfectly uniform and circular motions should be admitted. When Copernicus managed to eliminate the equants from his computations, he was regarded by most sixteenth-century astronomers as a *restorer* of the ancient purity of astronomy: the greatest astronomer since Ptolemy. The mobility of the earth they could regard as a 'mathematical' hypothesis which need not claim to be conformable to nature but only helpful as a practical device. It is true that not long afterwards the trepidation turned out to be a spurious phenomenon; nevertheless, at that time it belonged to the generally accepted 'facts' of nature. To bring it within the 'Platonic' framework was therefore considered a great scientific achievement.

If we were to take the standpoint of 'modern' science, however, our attitude would be quite the reverse: we would ignore Books II–VI as obsolete, and concentrate our attention on the first book, of which, to us, the most salient features are the introduction of the heliocentric system of the universe and the daily and annual motions of the earth. This interest of the modern scientist in Book I is thus very selective; it pays little attention to the *physical* explanations Copernicus put forward as an alternative to Aristotelian physics that lay behind the geocentric astronomy. To modern science the more or less pythagorico-platonic arguments then adduced by Copernicus are as obsolete as the Aristotelian and Ptolemaic views they replaced.

The first book, moreover, attracts attention from historians of all kinds because of the controversies it created after the Galileo trials; the conflict with the literalistic interpretation of some biblical texts then caused a great stir, of which even outsiders are aware up to the present day. It should be realized, however, that Copernicus' astronomy was not based on better observations than the Ptolemaic. Moreover, his physics was not more modern; it merely advanced more or less Platonic ideas over and above prevailing Aristotelian tenets. It set arguments against other arguments, but no decisive fact could be adduced in support. Whereas (before Copernicus) the seafarers had convinced everybody by *observations* proving that Ptolemy's *geography* was wrong in many respects, Copernicus could not adduce similar proofs that the physical basis of Ptolemy's *astronomy* was also wrong. Copernicus' advocacy was just a great achievement in the *art* of astronomy; it did not add new data to the 'history of nature'.

In consequence, the publication of Copernicus' theses caused far less stir than did the appearance of a new star in Cassiopeia in 1572. Thanks to Tycho Brahé's demonstration, the astronomers were compelled to recognize that it belonged not to the 'changeable' sublunar region, but to the allegedly unalterable sphere of the fixed stars which now turned out to be liable to change. This discovery was indeed a severe blow to ancient Aristotelian physics: it overthrew one of the central dogmas – the immutability of the heavenly regions. Controversy over the location of the *nova* led to a vast number of publications *pro et contra* – a number considerably greater than that caused by the appearance of Copernicus' book in 1543. What was now at stake was a controversy not between adherents of rival theoretical systems, but between a 'system' and a 'hard fact' – and *facts* counted heavily in the sixteenth century.

By 1600 there was still no observational proof for the Copernican system. In 1609, however, Galileo, with the help of his new telescope, discovered first the satellites of Jupiter and soon afterwards, the phases of Venus. The Jupiter satellites demonstrated that a planet could have moons, and thus made it plausible that the Earth, having a moon, was also a planet. Within the geocentric framework it was thought impossible that Venus would go through all the phases, whereas in Copernicus' system this must be the case. It should be pointed out, however, that this fact also fitted into Tycho's system. It was these observed *facts* that helped to turn Galileo from a lukewarm adherent of the Copernican system into its zealous apologist.[11]

11   With Galileo the relation between reason and observation was rather ambiguous. He was no empiricist in the Baconian sense – perhaps even less so than Kepler (he never abandoned the circularity of the celestial motions). He seems to have put greater trust in the reliability of human reason than in that of the human senses. He praised Aristarchus and Copernicus for having maintained that Venus revolves around the sun, although its apparent size seemed to remain the same: with them, according to Galileo, reason so much overpowered the senses that they made it 'the mistress of what they believed' and made the choice that has turned out to be the right one, as is now shown by the telescopic observations of the changing size and the phases of that planet. 'Nor can I sufficiently admire the eminence of these mens' wits that . . . have been able to prefer that which their reason dictated to them, to that which sensible experiments represented most manifestly on the contrary' (*Dialogue* III). Nevertheless, he was immensely proud of 'the perfection of our sight' by the invention of the telescope, for he, too, finally recognized that Experience is 'the true Mistress of Astronomy' (ibid.)

In many cases, better methods of observation confirmed his trust in Reason, but in his account of ebb and flood [1616; *Dialogue* IV (1633)] it led him astray. He selected experimental data that seemed to favour his mechanistic theory of the tides, according to which the basic phenomenon recurred in periods of twelve (instead of six hours). He ignored the generally known relation

Tycho Brahé's very precise observations of planetary motions were the basis of Kepler's *'Astronomia Nova sive Physica Coelestis'* (1609), a book that indeed inaugurated a new epoch in astronomical science. It was a mere difference of eight minutes which, in Kepler's own words, 'paved the way for the reformation of the whole astronomy'. Kepler now discarded all 'hypotheses' of excentres and epicycles from astronomy by introducing elliptic orbits. This, too, was an 'absurdity' – to Copernicans and Aristotelians alike – but the abolition of the epicycles was such a simplification that, in spite of the 'laws' of uniformity and circularity, ellipticity of orbits was gradually accepted as a physical fact. Kepler also made a first attempt at a mechanical explanation of the planetary motions (although he always wavered between the organistic and the mechanistic conceptions of the universe).

'Modern' astronomy, therefore, really began with Kepler who, though vainly searching for a satisfactory synthesis of physics and astronomy, nevertheless managed (on the basis of Tycho's exact measurements) to enunciate his three famous laws, which Newton later inserted into his system of cosmic physics. Taking together Kepler's astronomical discoveries and Galileo's telescopic observations, with their strong convincing power for many contemporaries, we may conclude that these two great scientists did indeed originate a new astronomy, based on new facts – an astronomy henceforth tending to go with a new mechanistic philosophy.

Historians of science who apply the evolutionistic method will tend to select the motions of the earth and the quasi-heliocentric structure of our planetary system as the features of the Copernican system that really matter. Still part and parcel of science, these enable us to trace a genealogical line from the initiator Copernicus to Newton, showing no deviation from the right path of progress. It is understandable that they then have some difficulty in evaluating Tycho, who gets good marks only for the reliability of his measurements, while his system is criticized as a regrettable step backwards. Although these historians may deplore or ignore the 'remnants' of Pythagorean and Aristotelian philosophies still extant in the works of Copernicus and some of his disciples, they easily pass them over as irrelevant to the progress of science.

If, on the other hand, a more phenomenological standpoint is taken, the re-enactment in our imagination will make us share the admiration of Copernicus' contemporaries for his restoring the self-consistency of the astronomical theory (eliminating equants and explaining trepidation) and we will see him as a representative of those humanists who sought the progress of science by a restoration of forgotten truths rather than by a revolution –

between the moon and the tides. According to his erroneous theory, ebb and flood could not occur if the earth were immovable. Consequently, he considered his theory of the tides one of the three main proofs of the Copernican system (the other two being the revolution of the sun around its axis and the retrogradations of the planets).

and also as a representative of that current of thought among Renaissance scholars which showed a predilection for Pythagorean and Platonic ideas.

With hindsight, we know that Copernicus' system did indeed provide the basis for Kepler and Galileo, but we should not forget that he himself was on the other side of the great watershed: his physics was organistic and not mechanistic. The great discoveries that undermined the Aristotelian philosophy of nature we owe not to him but to Tycho, and Galileo, whose telescopic observations greatly favoured the cause of heliocentric astronomy.

Francis Bacon and the New Philosophy

We have now to consider what the generations that came after the rise of sixteenth-century empiricism, who had witnessed the triumph of mechanistic philosophy, had to say about our problem.

The new philosophy advanced by the engineers, navigators and physicians, as well as by some philosophers, had sometimes been quite emphatically proclaimed; but more often it had been dispersed throughout their works in stray remarks and descriptions of experiments. It was finally put together in an eloquently worded programme by Francis Bacon (1561–1626). He contrasted the science of the future with that of past ages, and he did so in a quasi-biblical language that easily stuck in the minds of his contemporaries. Bacon was not very generous in mentioning the names of the sixteenth-century innovators he followed; but he formulated more elegantly and more systematically the ideas they had almost unwittingly and naively advanced, in a 'rude' style. Consequently, in the seventeenth century the names of his less sophisticated sixteenth-century predecessors remained in the shadow, and Bacon became the great prophet of the new natural history and the new experimental science.

Some of the great mathematico-physical scientists of the seventeenth century, who had themselves a considerable share in the mechanization of the world picture, praised Bacon highly.[12] Robert Boyle (1627–1691) was a thoroughgoing Baconian in his general approach; he called Bacon 'the great architect of experimental History'. Robert Hooke (1635–1703) larded his works with Baconian phrases, and although an outstanding mathematizing physicist, he agreed with Bacon that the 'history of nature and the arts' is the basis of science. Even Christiaan Huygens (1629–1695), the greatest of the mathematical physicists between Galileo and Newton, deemed Bacon the founder of a better 'philosophy', namely that which starts by experiments. Huygens and Leibniz agreed that experiments should be discussed methodi-

---

12   Cf. R. Hooykaas, 'De Baconiaanse Traditie in de Natuurwetenschap', *Algemeen Ned. Tijdschr. v. Wijsbegeerte* (1961), 53, pp. 181–201.

cally according to Bacon's plan, although both deplored his lack of mathematical knowledge.

The founders of the Royal Society (Boyle, Wallis, etc.) and of the Académie des Sciences started under the Baconian banner. Huygens, reporting at Colbert's request on the plan for the new academy, advised that it should mainly occupy itself with a 'natural history' of observations and experiments 'according to the plan of Verulamius (i.e., Bacon of Verulam).

Newton, although mentioning neither the rationalist Descartes nor the (rational-)empiricist Bacon, in his *'Principia'* (1687) took the side of the latter, as became explicit in Roger Cotes's preface to the second edition. The editor of the third edition, Henry Pemberton, wrote that Bacon was the first to combat speculative science and that he founded the true method of investigation of nature. Having expatiated on Bacon's principles Pemberton went on to expound Newton's methodological principles and showed their congruity with those of Bacon.[13] In the fierce battle of the Newtonians against Cartesian rationalism, Bacon's name came up again and again as that of the founder of 'experimental philosophy'.

The seventeenth-century founders of physico-mathematical science, therefore, regarded Bacon, in spite of his non-mathematical approach, as *the* great pioneer. But what did Bacon in fact bequeath to them, and – even more important for the problem we are now dealing with – what did Bacon himself think about the origin of the new science?

Bacon stressed above all other factors the fundamental role of *experience* over against speculative preconceptions. He warned his readers not to take 'authority for truth, instead of truth for authority'.[14] Like the sixteenth-century voyagers and the theological voluntarists, he insisted that facts must be accepted, however much they might seem to be against reason. Humility of spirit, obedience to the revelation in nature, he deemed indispensable to the investigator of nature. A philosophical conversion similar to a religious conversion was needed. This analogy of Bacon's is picked up by the late Benjamin Farrington, a well-known marxist historian of science and classical scholar, who succinctly expressed it in the heading of a chapter of his book on some of Bacon's minor works: 'Out with Aristotle, in with the Bible'.[15]

'What has been touched and seen' has to be accepted in spite of all rational prejudices; this tenet of the old navigators was also Bacon's: 'The entrance into the kingdom of Man, founded on the sciences, is not very different from the entrance into the kingdom of Heaven, whereinto none may enter except as a little child'.[16]

13   H. Pemberton, *A View of Sir Isaac Newton's Philosophy*, London, 1728, pp. 5 ff.

14   F. Bacon, 'Historia naturalis et experimentalis', in: *The Works of Francis Bacon* (eds J. Spedding, R. L. Ellis and D. D. Heath), London, 1857–1874, vol. II, p. 14.

15   B. Farrington, *The Philosophy of Francis Bacon*, Liverpool, 1964, p. 21.

16   Bacon, *Novum Organum I*, aph. 68; *Works I*, p. 179.

Secondly, Bacon, who realized how the great extension of knowledge caused by the voyages of discovery had exposed the old science as incomplete and often erroneous, wanted the development of a new Natural History to be the beginning of a scientific revolution.

Thirdly, for Bacon a new natural history should lead to a new natural philosophy. According to him, those who had handled science hitherto were either 'men of experience [solely]', or 'men of dogmas'.[17] The former he compared with the ants, which only collect extrinsic materials; the latter, the 'reasoners', with spiders which make cobwebs out of their own substance. But the true business of philosophy, so Bacon argues, is to act like the bee, which gathers material from the flowers and then digests and transforms it by a power of its own. The true philosophy neither relies chief on the powers of the mind, nor does it just collect the data provided by 'natural history and mechanical experiments'.

Fourthly, Bacon set natural history and 'mechanical experiments' on the same level. For him experiments are 'nature coerced by arts'; they yield reliable information, because even in *artificial* experiments we have to follow *nature*, which 'we cannot conquer, except by obeying her'. 'Therefore, from the closet and purer league between these two faculties, the experimental and the rational . . . much may be hoped'.

Apart from the general principles of the new science, Bacon tried also to give more precise methodological rules for 'digesting' the facts of natural and experimental history. He did not expect everybody to apply his rules, and he was open-minded enough to admit that the general principles of empiricism' and the art of experimentation could also be put into practice by adherents of the old school, or of new systems other than his own: '. . . when a true and copious history of nature and the arts shall once have been collected and digested . . . those great wits I spoke of before . . . will raise much more solid structures, and that too though they may prefer to walk on the old path, and not by way of my *Organum*, which in my estimation if not the only is at least the best course . . . My *Organum*, even if it were completed, would not without Natural History much advance the Instauration of the Sciences, whereas the Natural History without the *Organum* would advance it not a little.'[18]

Bacon's seventeenth-century disciples used the liberty granted to them: they wisely ignored most of his rules of inductive philosophy (though – also wisely – they did apply some of them). Practical men are, in general, too pragmatic to let themselves be shut up in a system, all the more so if this system shows signs of the limitations of an outsider who, although he delineates a marvellous general programme, does not know the whimsical tricks of nature from his own experience.

17   Bacon, *Nov. Org. I*, aph. 95; *Works I*, p. 201.
18   Bacon, *Historia Nat. et Exp.: Works II*, p. 15.

Bacon was at his best when contrasting the 'new' philosophy' with the old. Natural philosophy, said he, had been tainted either by logic (Aristotle) or by natural theology (Plato), or by mathematics (Proclus). Mathematics, in Bacon's opinion, should serve only to give definitions to natural philosophy, not to generate it.[19] This means that mathematics is useful for precise determination and measurement (i.e., for scientific 'description'), but that it ought not to be the basis of science. Evidently he was here rejecting the then widespread meta-mathematical, neopythagorean speculations about the ontological value of numbers and figures.

Bacon on the Voyages of Discovery

Bacon was firmly convinced that the voyages of discovery had coincided with the beginnings of the new natural history, and that the latter inevitably had to be followed by a new philosophy (i.e., science): '. . . by the distant voyages and travels which have become frequent in our times, many things have been laid open and discovered which may let in new light upon philosophy. And surely it would be disgraceful if, while the regions of the material globe – that is, of the earth, of the sea, and of the stars – have been in our time laid widely open and revealed, the intellectual globe should remain shut up within the narrow limits of old discoveries'.[20]

The opening up of the geographical globe by the voyages of discovery clearly caused the opening up of the intellectual globe, the new science and the new technological achievements: 'And this proficience in navigation and discovery may plant also great expectations of the further proficience and augmentation of the sciences; especially as it may seem that these two are ordained by God to be coevals, that is to meet in one age. For so the prophet Daniel [Dan. XII, 41], in speaking of the latter times, foretells: "That many shall go to and fro on the earth, and knowledge shall be increased", as if the opening and thorough passage of the world, and the increase of knowledge, were appointed to be in the same age; as we see it is already performed in great part'.[21]

Bacon's *religious* interpretation of the coincidence of the two events, as divinely pre-ordained, goes together with a *natural* explanation: there is a causal relation between the voyages and the ensuing astounding increase of knowledge of the 'history of nature and the arts', and the rise of new philosophy; the voyages are 'the causes and beginnings of great events'.

19   Bacon, *Nov. Org. I*, aph. 96; *Works I*, p. 201.
20   Bacon, *Nov. Org. I*, aph. 84; *Works I*, p. 191.
21   Bacon, *De Augmentis*, lib. II, c. 10; *Works I*, p. 514. Also: *Nov. Org. I*, aph. 93; *Works I*, p. 200.

Bacon clearly recognized, of course, that the voyages of discovery could not all of a sudden bring forth a new scientific system. But this process, he claims, has already begun though it seems hardly perceptible, and it will go on: 'This beginning was from God, the Father of Lights . . . Now in divine operations even the smallest beginnings lead of a certainty to their end. And as it was said of spiritual things, "The kingdom of God cometh not with observation", so it is in all the greater works of Divine Providence; everything glides so smoothly and noiselessly, and the work is fairly going on before men are aware that it has begun'.[22] It could hardly have been stated more clearly: the voyages of discovery (which drew everybody's attention!) were giving rise to a new science, initially almost imperceptible (or at any rate not yet perceptible by the multitude).

In Bacon's view – which we think was correct – the rise of the new science was not marked by a spectacular singular event. With him there are no stories like that of a stone dropped from Pisa's leaning tower, or an apple falling from a tree, or Haüy's calcspar crystal slipping from a visitor's hand. He does not offer material for hero-worshippers and hagiographers. On the contrary, he sees the rise of the new science as a general and gradual change of the intellectual climate, a change of method; and secondly, a change of world picture not restricted to one particular science (e.g., as astronomy), but affecting all scientific disciplines.

Conclusions

The rise of modern science had two major causes: firstly, the new natural history and the methodological epistemological changes connected with it; and secondly, the transition from an organistic to a mechanistic view of the world, a change closely connected with experimental philosophy and the contribution made to it by engineers, physicians, alchemists, cartographers, pilots and instrumentmakers.

Without any doubt, the view of nature held by modern science is mechanistic, so that 'mechanization' is one of the characteristic features of its rise. The term should, however, be taken in a wider sense than that of the mathematical formulation of the laws of statics, kinematics and dynamics. It also implies the use of mechanical (non-natural; artificial) instruments for the investigation of nature, the effacing of any radical distinction between the natural and the artificial, and the introduction of mechanical models of natural things.

22   Bacon, *Nov. Org. I*, aph. 95; *Works I*, p. 200.

Moreover, the new science is something cultivated by 'mechanicians' – mechanicians of the learned, liberal professions (physicians) as well as mechanicians in the proper sense of cultivators of illiberal arts (engineers, artisans, navigators). 'Mechanization' refers not only to a theory but also to a method; in a wider sense it embraces the contents (the substance) of science (nature as a mechanism, mechanistic philosophy) as well as its method (experimental philosophy). But perhaps the epistemological aspect of the new science is even more general. In natural sciences there were always rather reliable relations between Reason and Experience. Both of them have always been recognized as indispensable for the sound advancement of science, which has to steer through the narrow thoroughfare of rational empiricism, avoiding the rocks of both rationalism and naive empiricism.[23] Now Science, as a description and systematization of the facts given in nature, is a product of Reason: not of a sovereign, 'free' Reason, but of a Reason bound to 'data' and 'facta'. In physics, says Pascal, 'experience has a greater convincing power than reason': for we have to deal with nature, which remains just the same, regardless of the opinions we foster about her. In Science, therefore, the facts are the basis and touchstone of the theories. The great change (not only in astronomy or physics, but in *all* scientific disciplines) occurred when, not incidentally but in principle and in practice, the scientists definitively recognized the priority of Experience. The change of attitude caused by the voyages of discovery is a landmark affecting not only geography and cartography, but the whole of 'natural history'. It led to a reform of all scientific disciplines – (not only of the mathermatical – physical) – because it influenced the *method* of all the sciences, however much their mathematization might be delayed (as was the case, for example, with chemistry).

In discussing the rise of modern science, our educational past often influences our choice when deciding whether to lay emphasis upon the 'mechanization' or upon the new 'natural history'. Cultivators of the so-called 'exact' sciences will tend to concentrate attention on the rise of the new mechanics, together with the new astronomy. For them, therefore, the scientific 'revolution' begins with Galilean mechanics and ends with Newton's synthesis between the new astronomy and the new mechanics.[24] Copernicus' position

---

23   See fn. 11, on Galileo.
24   The late Professor E. J. Dijksterhuis (*De Mechanisering van het Wereldbeeld*, Amsterdam, 1950, pp. 319–332) allowed the period of the building up of the modern mechanistic world picture to run sharply from 1543 (Copernicus) to 1687 (Newton) (op. cit., p. 317). Further on, however, he said of Copernicus's work that 'apart from the use of the trigonometrical modes of computation, there is nothing in it that could not have been written in the 2nd century A.D. by a successor of Ptolemy' (op. cit., p. 319). Dijksterhuis's outline of Copernicus's theory is not essentially different from that we have given now. Moreover, he was an outspoken advocate of what he termed the 'phenomenological method' in historiography of science (cf. His *Doel en Methode* mentioned in fn. 2).

in the series of creators of classical – modern sciences then becomes ambiguous. With necessary reservations, one can draw some analogy between the relation of Kepler and Galileo to Copernicus and that of Lenin to Marx: both Copernicus and Marx put forward some fundamental ideas whose practical application did not come until six decades later.

Our choice of starting point may perhaps be determined even more by the historiographical method we apply, in particular when deciding whether we prefer to stress the importance of the geographical revolution or that of Copernicus in astronomy.[25] In the former case we will obviously emphasize the new natural history (geographical discoveries; revival of descriptive botany; observation of a new star and eventually Galileo's telescopic discoveries). Both parties, however, have to agree that it was not until 1600 that the sudden outburst of mechanistic philosophy and the astronomical reforms by Kepler and Galileo inaugurated the new astronomy and the new mechanics.

The considerable time lag between the earliest Portuguese oceanic voyages and the work of the early modern seventeenth-century scientists was an incubation period, in which the 'new philosophy' had already arisen, albeit almost noiselessly. In 1600, Gilbert published the results of research on magnetism performed in the past century (his own experiments included) under the title *Physiologia Nova*; and Kepler (1609) called his main work *Astronomia Nova*. Long before them, however (1513), a series of '*Tabulae Modernae*', based on the recent voyages of discovery, was added to Ptolemy's *Geographia* by its editor Waldseemuller. The 'geographical revolution' had preceded them by a whole century.

Henry the Navigator, who organized the first great voyages of discovery, was no scientist, and he had no scientific aims. But it was his initiative that

He, too, was of the opinion that modern astronomy really began with Kepler's *New Astronomy*: 'here we are confronted with one of the most important events in the history of thinking, perhaps even the real turning-point of the innovation that forms the theme of this book' (op. cit., p. 338). It hardly needs mentioning that Dijksterhuis found neither new important facts nor traces of a mechanistic world picture in the works of the astronomer, whom he nevertheless highly admired. (The title of Professor H. F. Cohen's recent inaugural address at the Technical University of Twente, *On the Character and Causes of the 17th Century Scientific Revolution* (Amsterdam, 1983), implies that in his opinion the 'revolution' did not start with Copernicus. His lecture develops a plan for a thorough investigation of the present topic.)

25   In particular, historians of science who have been educated as mathematicians, astronomers or physicists will have an open eye to the fact that physical (or mechanical) processes form the basis of all change in nature, and thus physics (or mechanics) is the most fundamental discipline. But not all sciences of nature have as yet been mathematized (or 'mechanicized'), although, nevertheless, they may claim to be 'scientific': empirical knowledge and classification are also 'science'. Many scientific discoveries, e.g., in chemistry, have been made without mathematization or mechanization (see: *Het Begrip Element – The Concept of Element*, pp. 145–159); this is even more so in botany and zoology. On the other hand, *all* sciences of nature are based on 'natural history': we start from facts and we end with facts which we classify, either in a mathematical or in a non-mathematical way.

triggered off[26] a movement which, growing into the avalanche of upheaval in sixteenth-century geography, opened the way for the reform, sooner or later, of all other scientific disciplines.

26   Aristotle made a distinction between the cause of a 'motion' (the transition from potentiality to actuality) and the incidental so-called 'cause' which is nothing but the removal of an obstacle hindering the true cause of nature ('. . . if anyone removes the obstacle he may be said in one sense – but in another not – to cause the movement; e.g. if he removes a column from beneath the weight it was supporting . . . for he accidentally determines the moment at which the potential motion becomes actual'. *Physica VIII*, 4; 255b, 20 ff).

    The physicist Robert Mayer, in an article 'Ueber Auslösung' (1876), spoke of 'loosening' (untying) or 'releasing' causes, in which there is no proportionality between cause and effect: a very small 'Anstoss' will, in general, have a much greater effect, e.g. when a light pressure of the finger 'causes' the enormous effect of a gun. He distinguished such release-causes from those about which he posited the thesis that 'the cause is equal to the effect', which he applied in his law of conservation of energy (R. Mayer, *Die Mechanik der Wärme* (ed. J. J. Weyrauch), Stuttgart, 1893, pp. 400–447). Such 'amplifying' processes are of course the basis of modern information technology.

2

Competing Disciplines

The Copernicans and the Churches

Robert S. Westman

Originally appeared as "The Copernicans and the Churches," in *God and Nature: Historical Essays on the Encounter between Christianity and Science*, edited by David C. Lindberg and Ronald L. Numbers (Berkeley, Los Angeles, and London: University of California Press, 1986): 76–113.

The text and notes in this chapter have been abridged. For complete footnotes the original publication should be referred to.

Editor's Introduction

Perhaps the most persistent stereotype about the interaction of science and religion is the "conflict model" which holds that science and religion are permanently and inevitably in opposition. It is not surprising that many students come to the history of science sharing this preconception almost unconsciously, since we constantly see it utilized in the modern media. It is also widely accepted and repeated by practicing scientists. Historians of science, however, rejected this stereotype long ago.

Of course there have been episodes where by appealing to revealed truths representatives of particular religions insisted that certain claims about nature were not only false but dangerous. The reception of Copernicus's heliocentric theory, culminating in the trial of Galileo, has been repeatedly deployed as the classic example of this. But if the conflict between Christianity and science is inevitable, why was there no condemnation of Copernicus for more than seventy years after the appearance of his work? It may surprise those who approach the Galileo affair from the perspective of the conflict thesis that Copernicus's heliocentric world system made hardly a splash in the sixteenth century. Neither the Catholic nor Protestant confessions expressly condemned it.

Rather, we understand such episodes better by localizing them; instead of reducing them to mere recurrences of some universal conflict we need to examine the particular details that make them unique. Here Robert Westman, an authority on the dissemination of Copernicanism, suggests we should regard the Galileo affair as the result of struggles among various disciplines. In every period of major scientific change, the boundaries between disciplines are challenged and redrawn. The early modern period was no exception. Westman argues that Copernicus's work was radical not primarily because of its heliocentrism, but because it violated the traditional hierarchy of disciplines. Traditionally, as a branch of mathematics, astronomy merely described and predicted celestial motions. It could not say what heavenly bodies actually were. This was the task of the higher discipline of natural philosophy. Copernicus claimed he was writing mathematics for mathematicians. His DE REVOLUTIONIBUS ORBIUM COELESTIUM (1543) was a very large and technical work of mathematical astronomy that very few people could actually read and understand. Most of those who could, primarily skilled mathematical practitioners, saw its main achievement in offering a way of eliminating the annoying equant, a mathematical construct which helped them make accurate predictions but violated the ancient dictum of perfect circular motion in the heavens. When Copernicus was read for his claims about the structure of the heavens, he was dismissed as a deluded trespasser on the domain of natural philosophy.

The other discipline involved in this story was theology. In the middle of the sixteenth century at the Council of Trent, the Catholic Church stated that only the Church and its theologians had the authority to interpret scripture. Unfortunately for Copernicans such as Galileo, many of the Church's theologians interpreted the Bible as saying the earth stood still. Whereas Copernicus violated disciplinary boundaries by making physical claims about the world based on mathematics, Galileo as a natural philosopher and mathematician violated them by claiming the ability to interpret scripture.

The Copernicans and the Churches

Robert S. Westman

In 1543, on his deathbed, Nicolaus Copernicus received the published results of his life's main work, a book magisterially entitled *De Revolutionibus Orbium Coelestium Libri Sex* (*Six Books on the Revolutions of the Celestial Orbs*), which urged the principal thesis that the earth is a planet revolving about a motionless central sun.[1] In 1616, seventy-three years after its author's death, the book was placed on the Catholic Index of Prohibited Books with instructions that it not be read "until corrected." Sixteen years later – and, by then, ninety years after Copernicus first set forth his views – Galileo Galilei (1564–1642) was condemned by a tribunal of the Inquisition for "teaching, holding, and defending" the Copernican theory. These facts are well known, but the dramatic events that befell Galileo in the period 1616–1632 have tended to overshadow the relations between pre-Galilean Copernicans and the Christian churches and to suggest, sometimes by implication, that the Galileo affair was the consummation of a long-standing conflict between science and Christianity.[2]

In this chapter we shall focus our attention on the long period between the appearance of *De Revolutionibus* and the decree of 1616. It will be helpful if we can suspend polar categories customarily used to describe the events of this period, such as Copernican versus anti-Copernican, Protestant versus Catholic, the individual versus the church. The central issue is better expressed as a conflict over the standards to be applied to the interpretation of texts, for this was a problem common to astronomers, natural philosophers, and theologians of whatever confessional stripe. In the case of the Bible, should its words and sentences in all instances be taken to *mean* literally what they say and, for that reason, to describe actual events and physical truths? Is the subject matter of the biblical text *always* conveyed by the literal or historical meaning of its words? Where does the ultimate authority reside to decide on the mode of interpretation appropriate to a given passage? In the case of an astronomical text, should its diagrams be taken to refer literally to actual paths of bodies in space? Given two different interpretations of the same celestial

---

1 Nuremberg, 1543; 2d edn, Basel, 1566. For English translations see A. M. Duncan, *Copernicus: On the Revolutions of the Heavenly Spheres* (Newton Abbot: David & Charles; New York: Barnes & Noble, 1976); and Edward Rosen, *Nicholas Copernicus: On the Revolutions*, vol. 2 of *Complete Works* (Warsaw and Cracow: Polish Scientific Publishers, 1978). All citations are to the Rosen translation. [. . .]

2 The *locus classicus* of this view is Andrew Dickson White, *History of the Warfare of Science with Theology in Christendom*, 2 vols. (New York: D. Appleton Century Co., 1936), 1:126. [. . .]

event, where does the authority reside to decide on the particular mode of interpretation that would render one hypothesis preferable to another? When the subject matters of two different *kinds of text* (e.g., astronomical and biblical or astronomical and physical) coincide, which standards of meaning and truth should govern their assessment? And finally, how did different accounts of the God-Nature relationship affect appraisal of the Copernican theory? Questions of this sort define the issue faced by sixteenth- and early seventeenth-century Copernicans.

Copernicus's Achievement

Before proceeding further, we must ask who Copernicus was and what he proposed. Nicolaus Copernicus (1473–1543) was a church administrator in the bishopric of Lukas Watzenrode, located in the region of Warmia, now northern Poland but then part of the Prussian Estates.[3] Watzenrode was Copernicus's uncle and guardian, and it was through his patronage that the young man was able to study medicine and canon law in Italy before returning to take up practical duties, including supervision of financial transactions, allocation of grain and livestock in peasant villages, and overseeing the castle and town defenses in Olsztyn. Though a member of the bishop's palace, Copernicus was not a priest but a clerical administrator or canon.

In his spare time Copernicus worried about a problem that had long concerned the church – accurate prediction of the occurrence of holy days such as Easter and Christmas.[4] Now calendar reform was an astronomical problem that demanded not primarily new observations but the assimilation of old ones into a model capable of accurately predicting the equinoxes and solstices, the moments when the sun's shadows produce days of longest, shortest, and equal extent. But predictive accuracy had never been the astronomer's only goal. The mathematical part of astronomy was complemented by a physical part.[5] The object of the latter was to explain why the planets moved, what they were made of, and why they are spaced as they are.[6] According to Aristotle's heavenly physics, the sun, moon, and other planets are embedded in great spheres made

3 On the life of Copernicus see Leopold Prowe, *Nicolaus Coppernicus*, 3 vols. (Berlin: Weidemann, 1883–1884); Ludwik Antoni Birkenmajer, *Mikolaj Kopernik* (Cracow: Uniwersytet Jagiellonskiego, 1900); and Edward Rosen, "Nicholas Copernicus: A Biography," in *Three Copernican Treatises*, 3d rev. edn (New York: Octagon Books, 1971), pp. 313–408.
4 See Noel Swerdlow, "On Copernicus' Theory of Precession," in *The Copernican Achievement*, ed. Robert S. Westman (Berkeley, Los Angeles, London: Univ. of California Press, 1975), pp. 49–98.
5 Considerable controversy exists among historians about how the domains of mathematics and physics were defined by contemporaries. [. . .]
6 For an excellent survey of the central elements of medieval cosmology see Edward Grant, "Cosmology," in *Science in the Middle Ages*, ed. David C. Lindberg (Chicago: Univ. of Chicago Press, 1978), pp. 265–302.

of a perfect and invisible substance called aether. The spheres revolve uniformly on axes that all pass through the center of the universe. This model
yielded an appealing picture of the universe as a kind of celestial onion with
earth at the core; but it failed to explain why the planets vary in brightness.
As an alternative, the astronomer Ptolemy (fl. AD 150) used a mathematical
device according to which the planet moves uniformly about a small circle (the
epicycle) while the center of the epicycle moves uniformly about a larger circle
(the deferent). Such a model could account for variations in both speed and
brightness. Ptolemy also invented another device, however, called the
"equant." [. . .] Here the center of an epicycle revolves *nonuniformly* as viewed
both from the sphere's center and from the earth but *uniformly* as computed
from a noncentral point (situated as far from the center on one side as the earth
is on the other). As a predictive mechanism the equant is successful. But now
ask how it can be that the planet, like a bird or fish, "knows" how to navigate
uniformly in a circle about an off-center point while, simultaneously, flying
variably with respect to the center of the same sphere? In response to objections like this it was quite customary for astronomers in the universities to consider the planetary circles *separately* from the spheres in which they were
embedded.[7] This meant that conflict between the mathematical and physical
parts of astronomy could be avoided by not mixing the principles of the two
disciplines. If, however, an astronomer were determined to reconcile physical
and mathematical issues, it would be customary within the Aristotelian tradition (which prevailed within the universities) to defer to the physicist, for in
the generally accepted medieval hierarchy of the sciences, physics or natural
philosophy was superior to mathematics.[8]

Copernicus, like all great innovators, straddled the old world into which he
was born and the new one that he created. On the one hand he was a conservative reformer who sought to reconcile natural philosophy and mathematical astronomy by proclaiming the absolute principle that all motions are
uniform and circular, with all spheres turning uniformly about their own
centers.[9] But, far more radically, Copernicus argued for the earth's status as a
planet by appealing to arguments from the *mathematical part* of astronomy.[10]
In so doing he shifted the weight of evidence for the earth's planetary status

7   The nature of medieval and Renaissance objections to the equant are still much disputed by
historians. [. . .]
8   In the division of the sciences going back to the thirteenth-century Parisian Aristotelian,
Albertus Magnus, the mathematical sciences are conceived as inferior to physics. This tradition
of the *divisio scientiarum* was opposed by a Platonic tradition at Oxford that claimed preeminence
for mathematics in the whole of natural philosophy; see James A. Weisheipl, "Classification of the
Sciences in Medieval Thought," *Medieval Studies* 27 (1965): 54–90, esp. 82–84.
9   *De Revolutionibus* 1.4, pp. 10–11; 5.2, p. 240.
10   Ibid., preface, p. 5; 1.10, p. 22; cf. Westman, "Astronomer's Role," pp. 109–111.

to the lower discipline of geometry, thereby violating the traditional hierarchy of the disciplines. If anything can be called revolutionary in Copernicus's work, it was this mode of argument – this manner of challenging the central proposition of Aristotelian physics.[11]

We are now prepared to consider the general logical structure of Copernicus's argument. Briefly, it looks like this: *If* we posit that the earth has a rotational motion on its axis and an orbital motion around the sun, *then* (1) all known celestial phenomena can be accounted for as accurately as on the best Ptolemaic theories; (2) the annual component in the Ptolemaic models, an unexplained mirroring of the sun's motion, is eliminated; (3) the planets can be ordered by their increasing sidereal periods from the sun; and (4) the distances of the planets from the center of the universe can be calculated with respect to a "common measure," the earth-sun radius (a kind of celestial yardstick), which remains fixed as the absolute unit of reference. [. . .] Although they were certainly among the most important consequences, these four were not the only ones to follow from the assumption of terrestrial motion. However, from the viewpoint of the prevailing logic of demonstrative proof, found in Aristotle's *Posterior Analytics*, there was no *necessity* in the connection between the posited cause and the conclusions congruent with that cause. Thus, while Copernicus's premises certainly authorized the conclusions he drew, there was no guarantee that other premises might not be found, equally in accord with the conclusions. In short, Copernicus had provided a systematic, logical explanation of the known celestial phenomena, but in making the conclusions the grounds of his premises, he failed to win for his case the status of a demonstrative proof.[12]

Pre-Galilean Copernicans were thus faced with several serious problems. First, their central premise had the status of an assumed, unproven, and (to most people) absurd proposition. Second, whatever probability it possessed was drawn primarily from consequences in a lower discipline (geometry). Third, even granting the legitimacy of arguing for equivalent predictive accuracy with Ptolemy, the practical derivation of Copernicus's numerical parameters was highly problematic. Fourth, the Copernican system flagrantly contradicted a fundamental dictum of a higher discipline, physics – namely, that a simple body can have only one motion proper to it – for the earth both orbited the sun and rotated on its axis. And finally, it appeared to conflict with the

11   For an alternative interpretation, arguing that the sole burden of justification still rests in the domain of natural philosophy, see Jardine, "Copernican Orbs," pp. 183–189.

12   On the concept of demonstrative proof see Aristotle, *Posterior Analytics* 1.6.74b15 ff. Cf. Owen Bennett, *The Nature of Demonstrative Proof According to the Principles of Aristotle and St. Thomas Aquinas*, The Catholic University of America Philosophical Studies, vol. 75 (Washington, DC: Catholic Univ. of America Press, 1943), pp. 58–85. It should be observed that Aristotle's own system of the planets does not satisfy the conditions of demonstrative proof.

interpretations of another higher discipline, biblical theology – in particular, the literal exegesis of certain passages in the Old Testament.

Under the circumstances Copernicus resorted to a rhetorical strategy of upgrading the certitude available to "mathematicians" – by which he meant those who practiced the mathematical part of astronomy – while underplaying the authority of natural philosophy and theology to make judgments on the claims of mathematicians.[13] Final authority for interpreting his text, he said, rested with those who best understand its claims. Church fathers such as Lactantius had shown a capacity for error in astronomy and natural philosophy, as when Lactantius declared the earth to be flat. Theologians of this sort should stay away from a subject of which they are ignorant.

Copernicus's strategy of appealing to the autonomy and superiority of mathematical astronomy was undercut by a prefatory "Letter to the Reader" that appeared immediately after the title page of De Revolutionibus. That brief epistle bespeaks the extraordinary circumstances surrounding the publication of the book.[14] It was only at the very end of Copernicus's life that he was finally persuaded to publish his book – not by one of his fellow canons, some of whom were eager to see the manuscript in press, but by a young Protestant mathematics lecturer who had come to visit the old canon from the academic heart of the Lutheran Reformation, the University of Wittenberg. Georg Joachim Rheticus (1514–1574) was permitted by Copernicus to publish a preliminary version of the heliocentric theory (Narratio Prima, 1540) and also to attend to the eventual publication of De Revolutionibus. But Rheticus lacked the time to oversee the work and so entrusted it to a fellow Lutheran, Andreas Osiander (1498–1552). Osiander, without permission from either Copernicus or Rheticus, took it upon himself to add an unsigned prefatory "Letter" written in the third person singular. Upon reading the manuscript, Osiander had become convinced that Copernicus would be attacked by the "peripatetics and theologians" on the grounds that "the liberal arts, established long ago on a correct basis, should not be thrown into confusion." Osiander hoped to save Copernicus from a hostile reception by appealing to the old formula according to which astronomy is distinguished from higher disciplines, like philosophy, by its renunciation of physical truth or even probability. Rather, if it provides "a calculus consistent with the observations, that alone is enough." De Revolutionibus was thus to be regarded as a strictly mathematical-astronomical text unable to attain even "the semblance of the truth" available to philosophers; and both mathematicians and philosophers were incapable of stating "anything certain unless it has been divinely revealed to them."[15]

13   See Westman, "Astronomer's Role," pp. 107–116.
14   See A. Bruce Wrightsman, "Andreas Osiander's Contribution to the Copernican Achievement," in Copernican Achievement, ed. Westman, pp. 213–243.
15   De Revolutionibus, p. xvi. The modern editor and translator Edward Rosen has labeled the Letter "Foreword by Andreas Osiander," but no such designation appears in the original work.

Early Protestant Reaction: The Melanchthon Circle and the "Wittenberg Interpretation"

When Rheticus returned to his teaching duties at Wittenberg after his long visit to Copernicus, he brought back strongly favorable personal impressions of the Polish canon and his new theory.[16] Rheticus himself was Copernicus's first major disciple, and many of the Wittenberger's students read and studied *De Revolutionibus*. Furthermore, Rheticus composed a treatise, recently rediscovered, in which he sought to establish the compatibility of the Bible and the heliocentric theory.[17] All of this tempts us to ask whether Protestants were particularly well disposed toward the Copernican theory.

To answer this question, we must distinguish between the Protestant Reformers and men who happened to be Protestants and were also well versed in the reading of astronomical texts. The Reformers Luther and Calvin were learned men who knew enough astronomy to understand its basic principles; but neither had ever practiced the subject. It used to be thought that Luther played an important role in condemning Copernicus's theory when, in the course of one of his *Tischreden* or *Table Talks*, he said: "That fool wants to turn the whole art of astronomy upside down."[18] But the statement itself is vague on details and, in any event, was uttered in 1539, sometime before the publication of either Rheticus's *Narratio Prima* or Copernicus's *De Revolutionibus*. As for Calvin, there is no positive evidence that he had ever heard of Copernicus or his theory; if he knew of the new doctrine, he did not deem it of sufficient importance for public comment. In short, there are no known opinions by these two leading Protestant Reformers that significantly influenced the reception of the Copernican system.[19]

There was, however, a third Reformer, a close associate of Luther's and the educational arm of the Reformation in Germany, Philipp Melanchthon (1497–1560), known as *Praeceptor Germaniae*. A charismatic man, beloved

16 Karl Heinz Burmeister has published all of Rheticus's correspondence, a detailed bibliography of his works, and an excellent biography: *Georg Joachim Rhetikus, 1514–1574: Eine Bio-Bibliographie*, 3 vols. (Wiesbaden: Guido Pressler Verlag, 1967). For a different interpretation of Rheticus's relationship to Copernicus see Robert S. Westman, "The Melanchthon Circle, Rheticus and the Wittenberg Interpretation of the Copernican Theory," *Isis* 66 (1975): 165–193, esp. 181–190.

17 As this article goes to press, R. Hooykaas announces the discovery of an anonymous treatise entitled *Letter on the Motion of the Earth* (1651) that he believes to be Rheticus's ("Rheticus's Lost Treatise on Holy Scripture and the Motion of the Earth," *Journal for the History of Astronomy* 15 [1984]: 77–80).

18 Luther's words are reported here by Aurifaber: "Der Narr will die ganze Kunst Astronomiae umkehren" (*Weimar Ausgabe, Tischreden*, vol. 1, no. 855). [. . .]

19 On the Reformers' reaction to the Copernican theory see Richard Stauffer, "L'attitude des Réformateurs à l'égard de Copernic," in *Avant, avec, après Copernic: La représentation de l'univers et ses conséquences épistémologiques* (Paris: Albert Blanchard, 1975), pp. 159–163.

teacher, and talented humanist, Melanchthon was also a brilliant administrator with a gift for finding compromise positions.[20] In the face of serious disturbances from the Peasants' Revolt of 1524–1525 and plunging enrollments all over Germany, Melanchthon instituted far-reaching reforms that led to the rewriting of the constitutions of the leading German Protestant universities (Wittenberg, Tübingen, Leipzig, Frankfurt, Greifswald, Rostock, and Heidelberg), profoundly influencing the spirit of education at several newly founded institutions (Marburg, Königsberg, Jena, and Helmstedt). Most important of all, Melanchthon believed that mathematics (and thus astronomy) deserved a special place in the curriculum because through study of the heavens we come to appreciate the order and beauty of the divine creation. Furthermore, mathematics was an excellent subject for instilling mental discipline in students. Such views alone would not predispose one toward a particular cosmology, but they did help to give greater respectability to the astronomical enterprise. Thus, a powerful tradition of mathematical astronomy developed at Wittenberg from the late 1530s and spread throughout the German and Scandinavian universities. At Wittenberg itself, three astronomers in the humanistic circle gathered around Melanchthon were preeminent: Erasmus Reinhold (1511–1553), his pupil Rheticus, and their joint pupil and the future son-in-law of Melanchthon, Caspar Peucer (1525–1603). Melanchthon was the *pater* of this small *familia scholarium*. Many of the major elements in the subsequent interpretation of Copernicus's theory in the sixteenth century would be prefigured in this group at Wittenberg.

The "Wittenberg Interpretation," as we will call it, was a reflection of the views of the Melanchthon circle. Melanchthon himself was initially hostile to the Copernican theory but subsequently shifted his position, perhaps under the influence of Reinhold. Melanchthon rejected the earth's motion because it conflicted with a literal reading of certain biblical passages and with the Aristotelian doctrine of simple motion. But Copernicus's conservative reform – his effort to bring the calculating mechanisms of mathematical astronomy into agreement with the physical assumption of spheres uniformly revolving about their diametral axes – was warmly accepted. Reinhold's personal copy of *De Revolutionibus*, which still survives today, is testimony; it has written carefully across the title page the following formulation: "The Astronomical Axiom: Celestial motion is both uniform and circular or composed of uniform and circular motions." As it stands, this proposition simply ignores physical claims for the earth's motion, but commits itself to an equantless astronomy. It is, we might say today, a "research program," one which Copernicus tried to make compatible with the assumption that the earth is a planet. But the Wittenbergers, with the noticeable exception of Rheticus, refused to follow

20 On Melanchthon and the Wittenberg school see Westman, "Melanchthon Circle," pp. 165–181.

Copernicus in upsetting the traditional hierarchy of the disciplines. Instead, Reinhold and his extensive group of disciples accepted Melanchthon's physical and scriptural objections to the Copernican theory. In the prevalent mood of reform, Copernicus was perceived not as a revolutionary but as a moderate reformer (like Melanchthon), returning to an ancient, pristine wisdom before Ptolemy.

If Melanchthon and Reinhold saw Copernicus as a temperate reformer, Rheticus saw the radical character of his reform. Rheticus returned to Wittenberg in 1542 as an inflamed convert, writing of Copernicus as of one who has had a Platonic vision of The Good and The Beautiful – though in the harmony of the planetary motions. "My teacher," wrote Rheticus, referring to Copernicus,

> was especially influenced by the realization that the chief cause of all the uncertainty in astronomy was that the masters of this science (no offense is intended to divine Ptolemy, the father of astronomy) fashioned their theories and devices for correcting the motion of the heavenly bodies with too little regard for the rule which reminds us that the order and motions of the heavenly spheres agree in an absolute system. We fully grant these distinguished men their due honor, as we should. Nevertheless, we should have wished them, in establishing the harmony of the motions, to imitate the musicians who, when one string has either tightened or loosened, with great care and skill regulate and adjust the tones of all the other strings, until all together produced the desired harmony, and no dissonance is heard in any.[21]

Even more enthusiastically than Copernicus, Rheticus extolled the "remarkable symmetry and interconnection of the motions and spheres, as maintained by the assumption of the foregoing hypotheses," appealing to analogical concordance with musical harmonies, to the number six as a sacred number in Pythagorean prophecies, to the harmony of the political order in which the emperor, like the sun in the heavens, "need not hurry from city to city in order to perform the duty imposed on him by God," and to clockmakers who avoid inserting superfluous wheels into their mechanisms. Copernicus's unification of previously separate hypotheses had a liberating, almost intoxicating, effect on Rheticus, which Rheticus expressed almost as a personal revelation fully comprehensible only by visualizing the ideas themselves.

A wide spectrum of early Protestant opinion is defined between Melanchthon's cautious promotion of Copernicus's reform and Rheticus's radical espousal of the core propositions of Copernican cosmology. In general, the Wittenberg Interpretation dominated until the 1580s, while Rheticus's vision was typically ignored in public discussions. By the late 1570s, however, there were signs of the emergence of a cosmological pluralism among

21   Georg Joachim Rheticus, *Narratio Prima*, in Rosen, *Three Copernican Treatises*, p. 138.

Protestant astronomers.[22] A Danish aristocrat named Tycho Brahe (1546–1601) established an extraordinary astronomical castle on the misty island of Hveen, near Copenhagen, where he commenced a major reform of astronomical observations and, by the early 1580s, proposed a new cosmology in which all the planets encircle the sun, while the sun moves around the stationary, central earth. This system – the Tychonic or geoheliocentric – adopted Copernican-heliocentric paths for the planets, causing the orbits of Mars and the sun to intersect, while preserving Aristotelian terrestrial physics [. . .], but in another quite important respect, Tycho departed from Aristotle by abolishing the solid celestial spheres.[23] In 1600 the Englishman William Gilbert (1540–1603) suggested that the earth possesses a magnetic soul that causes it to turn daily on its axis; but he was cryptic about the ordering of Mercury and Venus.[24]

Throughout the second half of the sixteenth century, Copernicus's book was widely read and sometimes studied in both Catholic and Protestant countries.[25] Compared to the fairly large number of people aware of the central claims of De Revolutionibus, however, there were relatively few who actively adopted its radical proposals and whom we can justifiably call "Copernicans" in that sense. To be precise: we can identify only ten Copernicans between 1543 and 1600; of these, seven were Protestants, the others Catholic. Four were German (Rheticus, Michael Maestlin, Christopher Rothmann, and Johannes Kepler); the Italians and English contributed two each (Galileo and Giordano Bruno; Thomas Digges and Thomas Harriot); and the Spaniards and Dutch but one each (Diego de Zuñiga; Simon Stevin). It is time now to examine the Catholic reaction more closely.

Early Catholic Reaction and the Council of Trent

De Revolutionibus was published at a time when two powerful socioreligious movements converged. The first of these was the Protestant Reformation and, of special importance to us, its incursion into the German universities. The second was a movement of reform from within the Catholic church, the

22   See Robert S. Westman, "Three Responses to the Copernican Theory: Johannes Praetorius, Tycho Brahe and Michael Maestlin," in *Copernican Achievement*, ed. Westman, pp. 285–345; and Owen Gingerich, "Copernicus and Tycho," *Scientific American* 229, no. 6 (Dec. 1973), 86–101. [. . .]
23   In 1588, the same year in which Tycho's system was published, Nicholas Reymers Baer proposed a system without intersecting spheres that makes the planets revolve around both the sun and the earth while the latter rotates on its axis at the center of the universe (*Fundamentum Astronomicum*, Strassburg). A furious priority dispute ensued. For brief details see J. L. E. Dreyer, *Tycho Brahe* (Edinburgh, 1890; reprint, New York: Dover, 1963), pp. 183–185.
24   *De Magnete*, trans. P. Fleury Mottelay (London: Quaritch, 1893), 6.3.
25   Owen Gingerich, "The Great Copernicus Chase," *American Scholar* 49 (1979): 81–88.

Counter-Reformation, driven partly by the need to respond to the Protestant challenge, partly by the need of the papacy to assert its authority in an area where for too long it had avoided reform. In 1545 Pope Paul III (1534–1549), to whom Copernicus had dedicated *De Revolutionibus*, called into session a council in the Italian Imperial city of Trent, which was to last until 1563, the year before Galileo's birth. A list of laxities within the church can only hint at the depths of the need for reform: cardinals, bishops, and priests chronically absent from their domains of responsibility; irregularities in clerical training and abysmal literacy levels among parish priests; rampant granting of privileges and dispensations; priestly ownership of private land; and unchecked drunkenness, concubinage, and hunting among the clergy.[26] No wonder that the overriding issue confronting the Council was the need to give the faithful some feeling of security by restoring clerical discipline and providing a highly structured theology. Whatever real reforms were eventually made, however, the new initiatives created an atmosphere of obsessional control over detail, endless doctrinal clarifications by councils, synods, and theologians, suspicion of deviancy, and a proclivity for inflexible, legalistic remedies in areas of social conflict – a climate to which Protestants also contributed.[27]

At Trent the problem of authority surfaced in many ways, not least in the question of the authenticity of the Catholic Bible or Vulgate, the Latin translation prepared by Saint Jerome in the fourth century. After considerable debate it was agreed that the Vulgate, together with the writings of the church fathers, was to be the final authority in all matters of faith and discipline. [. . .]

Copernicus's theory was not discussed at the Council of Trent. In fact, matters of natural philosphy and even calendrical reform were not in any sense primary issues of discussion.[28] But now, thanks to recently discovered evidence, we know that about the time the Council was beginning, there was some considered reaction within certain circles at the papal court. A year after the appearance of *De Revolutionibus*, a Florentine Dominican theologian-astronomer, Giovanni Maria Tolosani (1470/1–1549), authored a large apologetic work entitled *On the Truth of Sacred Scripture*.[29] This treatise, completed in 1544 but never published, concerned itself precisely with certain

26 See "The Beginnings of the Catholic Reform in Rome under Paul III," in *History of the Church*, ed. Hubert Jedin and John Dolan, trans. Anselm Biggs and Peter Becker, 10 vols. (London: Burns & Oates, 1965–1980), 5:456–462.

27 See Jean Delumeau, *Catholicism between Luther and Voltaire: A New View of the Counter-Reformation*, trans. Jeremy Moiser (London: Burns & Oates Ltd., 1977; original French version published in 1971), p. 126.

28 This statement is based upon a careful study of the indices of *Concilium Tridentinum: Diariorum, Actorum, Epistularum, Tractatuum*, ed. Stephanus Ehses et al., 13 vols. (Freiburg im Breisgau: Herder, 1961–1976).

29 Firenze, Biblioteca Nazionale, MS Conv. Soppr. J.I.25, with the provenance of the monastery of San Marco.

issues that were about to be debated at the Council of Trent. Between 1544 and 1547 Tolosani added a cluster of "little works" dealing with such topics as the power of the pope and the authority of councils, conflict between Catholics and heretics, justification by faith and works, the dignity and office of cardinals, and the structure of the church. But Tolosani was also an astronomer of no mean ability. He had written a treatise on the reform of the calendar and had attended the Fifth Lateran Council (1515), where calendrical reform was a primary topic.[30] Somehow he obtained a copy of *De Revolutionibus* and wrote in his fourth *opusculum* an extensive critique of book 1.[31] This little work is remarkable. First, in contrast with the Wittenberg Interpretation, it avoided purely technical astronomical issues and made no mention of the equant. Second, Copernicus's theory conflicted with the most basic principles of Aristotelian natural philosophy. And third, the Dominican Tolosani chose to locate *De Revolutionibus* within the Thomist hierarchy of the disciplines and to present the theory as a violation of its principles of classification.

> He [Copernicus] is expert indeed in the sciences of mathematics and astronomy, but he is very deficient in the sciences of physics and dialectic. Moreover, it appears that he is unskilled with regard to Holy Scripture, since he contradicts several of its principles, not without the danger of infidelity to himself and to the readers of his book. . . . The lower science receives principles proved by the superior. Indeed, all the sciences are connected mutually with one another in such a way that the inferior needs the superior and they help one another. An astronomer cannot be perfect, in fact, unless first he has studied the physical sciences, since astrology [i.e., astronomy] presupposes celestial corporeal natures and the motions of these natures. A man cannot be a complete astronomer and philosopher unless through logic he knows how to distinguish between the true and the false in disputes and knows the modes of argumentation, [skills] that are required in the medicinal art, philosophy, theology, and the other sciences. Hence, since Copernicus does not understand physics and logic, it is not surprising that he should be mistaken in this opinion and accepts the false as true, through ignorance of these sciences. Call together men well read in the sciences, and let them read Copernicus's first book on the motion of the earth and the immobility of the sidereal heaven. Certainly they will find that his arguments have no force and can very easily be resolved. For it is stupid to contradict an opinion accepted by everyone over a very long time for the strongest reasons, unless the impugner uses more powerful and incontrovertible demonstrations

30   See Demetrio Marzi, *La questione della riforma del calendario nel Quinto Concilio Lateranense (1512–1517)* (Florence: G. Carnesecchi e Figli, 1896).

31   "De coelo supremo immobile et terra infima stabili ceterisque coelis et elementis intermediis mobilibus." The entire text of this little work is transcribed by Eugenio Garin, "Alle origini della polemica anticopernicana," in *Colloquia Copernicana*, vol. 2, Studia Copernicana, vol. 6 (Wroclaw: Ossolineum, 1975), pp. 31–42. [. . .]

and completely dissolves the opposed reasons. But he [Copernicus] does not do this in the least.[32]

What we see here is the possibility of exploiting the somewhat ambiguous status of astronomy as a mixed or middle science. Unlike Osiander, who tried to protect Copernicus by stressing the *separation* between the mathematical and the physical parts of astronomy, Tolosani brings out the *dependency* of astronomy upon the higher disciplines of physics and theology for the truth of its conclusions. Physics and theology are disciplines superior to mathematics by virtue of their sublime subject matters, tradition, and demonstrative capacity to reach necessary conclusions. The rejection of *De Revolutionibus* as inconsistent with a very conservative rendering of Aristotle's physics (failing even to mention Thomas Aquinas's notion of impressed force) was entirely consonant with the mood that was to prevail at Trent regarding the exegesis of biblical passages.

Tolosani ends his little treatise with the following interesting revelation: "The Master of the Sacred and Apostolic Palace had planned to condemn this book, but, prevented first by illness and then by death, he could not fulfill this intention. However, I have taken care to accomplish it in this little work for the purpose of preserving the truth to the common advantage of the Holy Church."[33] The Master of the Sacred Palace was Tolosani's powerful friend, Bartolomeo Spina, who attended the opening sessions of the Council of Trent but died in early 1547.[34] As trenchant as Tolosani's critique of Copernicus had been, there is simply no evidence that it received any serious consideration either from the new master or from the pope himself. Meanwhile, Tolosani's unpublished manuscript, written in the spirit of Trent, was probably shelved in the library of his order at San Marco in Florence awaiting its use by some new prosecutor. The result was that sixteenth-century Catholic astronomers and philosophers worked under no formal prohibitions from the Index or the Inquisition.

The Copernican Theory and Biblical Hermeneutics

The Protestant Reformers were agreed in emphasizing the plain, grammatical sense as the center of biblical interpretation, thereby making it accessible to anyone who could read. Additional help was sometimes sought from spiritual or allegorical readings, but the literal, realistic meaning always remained

---

32   Garin, "Alle origini," pp. 35–36.
33   Ibid., p. 42.
34   Spina presented articles concerning baptism and justification (*Concilium Tridentinum* 12:676, 725); he was succeeded on his death by the Bolognese Dominican Egidius Fuscharus (ibid., 5:728n).

central. Now, the literalism of the Reformers was twofold: they believed that the Bible was literal both at the level of direct linguistic reference (nouns referred to actual people and events) and in the sense that the *whole story* was realistic. The Bible's individual stories needed to be woven together into one cumulative "narrative web." This required the earlier stories of the Old Testament to be joined interpretatively to those in the New Testament by showing the former to be "types" or "figures" of the latter. Luther and Calvin were agreed that there was a single theme, a primary subject matter, which united all the biblical stories: the life and ministry of Christ.[35]

Although Protestants rejected the Catholic appeal to allegorical and ana-gogical interpretations of Scripture as an illegitimate stretching of the plain meaning, both groups of exegetes had available to them a method of inter-pretation to which they could appeal: the principle of accommodation. One purpose of this hermeneutic device was to resolve tensions between popular speech, wedded to the experience of immediate perception, and the specialized discourse of elites. The necessity of sacrifices or anthropomorphic references to God as a man with limbs were types of references that could easily evoke appeal to the principle of accommodation.[36] In the seventeenth and eighteenth centuries, Jesuit missionaries in China sparked a controversy over accommo-dation when they allowed Chinese converts to pray to Confucius, worship ancestors, and address God as *Tien* (sky).[37] Like the Jesuit missionaries, the sixteenth-century followers of Copernicus made use of the option of accom-modation. For them, however, the problem was not the alien belief-systems of a foreign society but the disciplinary hierarchy of the universities in which the-ology occupied the highest rank.

Before pursuing this matter further, let us look briefly at four specific classes of biblical passages that were relevant to the Copernican issue – references to the stability of the earth, the sun's motion with respect to the terrestrial horizon, the sun at rest, and the motion of the earth. Both Protestant and Catholic geocentrists customarily cited verses from the first two categories and interpreted them to refer literally to the physical world. Consider, for example, Psalms 93:1: "The world also is stablished, that it cannot be moved"; or Ecclesiastes 1:4: "One generation passeth away, and another generation cometh: but the earth abideth for ever"; Ecclesiastes 1:5: "The sun also

35   See Hans Frei, *The Eclipse of Biblical Narrative: A Study in Eighteenth and Nineteenth Century Hermeneutics* (New Haven: Yale Univ. Press, 1974), pp. 1–37.

36   On the problem of accommodation see Klaus Scholder, *Ursprunge und Probleme der Biblelkri-tik im 17. Jahrhundert* (Munich: Kaiser, 1966), pp. 56–78; Amos Funkenstein, "The Dialectical Preparation for Scientific Revolutions: On the Role of Hypothetical Reasoning in the Emergence of Copernican Astronomy and Galilean Mechanics," in *Copernican Achievement*, ed. Westman, pp. 195–197.

37   See Johannes Bettray, *Die Akkommodationsmethode des P. Matteo Ricci S.I. in China* (Rome: Universitas Gregoriana, 1955), p. 278.

ariseth, and the sun goeth down and hasteth to his place where he arose"; Psalm 104:19: "He appointed the moon for seasons: the sun knoweth his going down."[38] The literal interpretation of these passages springs from different sources for Protestants and Catholics. For Protestants, such as Melanchthon, it came from a steadfast faith in the inerrancy of the grammatically literal text; for Catholics, such as Tolosani, the literal meaning was legitimated by appeal to the (allegedly unanimous) authority of previous interpreters. In both cases the geocentrists ignored verses from categories three and four.

The Copernicans had available to them two hermeneutical strategies. The first, which we may call "absolute accommodationism," declares that the verses in all four categories are accommodated to human speech. The virtue of this position is that it draws a radical line of demarcation between biblical hermeneutics and natural philosophy, so that the principles and methods of the one cannot be mixed with those of the other. It is also in keeping with the moderate Christocentric reading of Scripture advocated by the Reformers. Far more dangerous was the second strategy, which we may call "partial accommodationism," according to which the interpreter provides a literal, *heliostatic* or *geomotive*, construal of either Joshua 10:12–13 or Job 9:6 and then accommodates it to verses conventionally read as geostatic. In the Joshua text we read: "Then spake Joshua to the Lord in the day when the Lord delivered up the Amorites before the children of Israel, and he said in the sight of Israel, Sun, stand thou still upon Gibeon; and thou, Moon, in the valley of Ajalon. And the sun stood still, and the moon stayed, until the people had avenged themselves upon their enemies." The construction "stand still" is certainly plain talk to the senses; thus, the heliocentrist, if he wished to pursue a partial-accommodationist line, must point out that we need not intend the horizon as our reference frame and that the sun could be rotating on its own axis, while remaining at rest at the center of the universe. A similar kind of ambiguity of reference frame is present in the Job text: "Which shaketh the earth out of her place, and the pillars thereof tremble." The phrases "out of her place" and "tremble" could be taken to denote either diurnal or annual motion or simply the earth quaking. The sixteenth-century Copernicans, perhaps taking the lead from Copernicus's brief remarks about Lactantius, tended to adopt the position of absolute accommodation. The two most eloquent expressions of this position were by Giordano Bruno (1548–1600) and Johannes Kepler (1571–1630).

Bruno, a Dominican from Naples, was well trained in Scholastic philosophy and Thomist theology, but he had also been receptive to newer, radical intellectual currents, including the cosmology of Copernicus.[39] In 1576 he

---

38   Another verse cited was Isaiah 66:1: "Thus saith the Lord, the heaven is my throne, and the earth is my footstool."

39   There is a considerable literature on Bruno. [. . .]

suddenly left his order in Naples and began a fifteen-year pilgrimage throughout Europe, preaching on a variety of subjects including the deficiencies of
Aristotle. Describing himself as an "academician of no academy," Bruno
arrived at Oxford in 1583, and there defended Copernicus's theory before a
hostile audience of philosophers and theologians. The following year he published a witty and sarcastic humanistic dialogue called *The Ash Wednesday
Supper*, where he dealt with the problem of Copernicus and the Bible. Referring to Ecclesiastes 1:5, he wrote ironically:

> So if the Sage, instead of saying, "The sun riseth and goeth down, turneth toward
> the south and boweth to the north wind," had said: "The earth turns round to
> the east, leaving behind the sun which sets, bow to the two tropics, that of Cancer
> to the south and Capricorn to the north wind," his listeners would have stopped
> to think: "What, does he say that the earth moves? What kinds of fables are
> these?" In the end, they would have accounted him a madman, and he really
> would have been a madman.[40]

Moral and redemptive meaning rather than physical truth was the Bible's
message, according to Bruno. [. . .]

There was one notable exception to this Copernican consensus on accommodation: Diego de Zuñiga (1536–1597). A student of Luis de León at
Salamanca and, like him, a member of the Hermits of St Augustine, Zuñiga
taught theology and philosophy at the universities of Osuna and Salamanca.[41]
Zuñiga's writings were Augustinian in theme: in 1577 a treatise on free
will, and in 1584 a commentary on the Book of Job.[42] Zuñiga's sentiments
were clearly with the liberal Hebraist faction at Salamanca, although he
had not been present at Trent and would not go as far as Fray Luis in approv-
ing vernacular translations. In his massive commentary on Job, Zuñiga had
to face the difficult problem of chapter 9, verse 6. The general theme of this
part of Job is the omnipotence and wisdom of God, who "shaketh the earth
out of her place, an the pillars thereof tremble." Perhaps wishing to emphasize God's physical and moral power, Zuñiga did not see the language as
accommodated to common speech but read it literally as a statement about
the physical world; in particular, he explicated the passage according to the
Copernican theory.[43] [. . .]

---

40   Giordano Bruno, *The Ash Wednesday Supper*, ed. and trans. Edward A. Gosselin and Lawrence
S. Lerner (Hamden, Conn.: The Shoe String Press, 1977), Dialogue 4, p. 178.
41   On Zuñiga's life and work see Marcial Solana, *Historia de la filosofia española*, 3 vols. (Madrid:
Aldus, 1940), 3:221–260. [. . .]
42   *De Vera Religione in Omnes Sui Temporis Haereticos Libri Tres* (Salamanca, 1577); *In Zacharias
Prophetam Commentaria* (Salamanca, 1577); *In Job Commentaria* (Toledo, 1584). Zuñiga's aim was
nothing less than a full commentary on all books of the Bible.
43   [. . .] An English translation of the Zuñiga passage was made by Thomas Salusbury in the
seventeenth century: *Mathematical Discourses and Demonstrations* (London, 1665), pp. 468–470.

The Jesuits

The Society of Jesus, founded at about the time the Council of Trent began and *De Revolutionibus* was published, was the real sword of the Counter-Reformation. Worldly and militant, the Jesuits eagerly engaged the Protestants in polemics and ingratiated themselves with the royal courts as privileged advisers. Even more impressively, they challenged the Melanchthonian hegemony by founding their own colleges all over Europe.[44] Systematically dividing the Continent into regions, they rapidly established dozens of colleges in the 1550s and 1560s within each area from the Iberian Peninsula to the Provinces of the Netherlands, from the German principalities east to the Hapsburg lands. The flagship of these colleges was the Collegio Romano in Rome. Its professors were among the best in the Society and were the leaders in establishing curricular policy for the college system. Perhaps because of this position of leadership, the Collegio Romano was also a site of controversy among its most important lecturers. Not only was wrangling prevalent within disciplines, but serious debates also occurred between the philosophers and mathematicians. In the 1580s, debates over educational policy came to a head with the promulgation of the *Ratio Studiorum*.[45] A primary author of this document, Christopher Clavius (1537–1612), succeeded in elevating the status of mathematics to an unprecedented level of academic responsibility, arguing for its pedagogical indispensability to philosophy and the other disciplines.[46]

Clavius, an outstanding astronomer and mathematician, disagreed particularly with those philosophers who had no practical experience as astronomers, yet insisted on questioning the physical reality of the mechanisms posited by Ptolemaists like Clavius.[47] The debate then turned on the degree of certitude to which astronomy could aspire in constructing true explanations. In his authoritative textbook, which became the standard of the Jesuits, Clavius argued that astronomy, like physics, was concerned with true causes. The Averroists at the Collegio Romano, who believed in solid,

---

44   See, for example, François de Dainville, S.J., "L'enseignment des mathématiques dans les collèges jesuites de France du XVIe au XVIIIe siècle," *Revue d'histoire des sciences et de leurs applications* 7 (1954): 6–21, 102–123.

45   The full text and translation of the relevant passages is given in A. C. Crombie, "Mathematics and Platonism in the Sixteenth Century Italian Universities and in Jesuit Educational Policy," in *Prismata: Naturwissenschaftsgeschichtliche Studien: Festschrift für Willy Hartner*, ed. Y. Maeyama and W. G. Saltzer (Wiesbaden: Franz Steiner Verlag, 1977), pp. 63–94, esp. 65–66. [. . .]

46   On Clavius, the Collegio Romano, and their importance for Galileo, the definitive work is now William A. Wallace, *Galileo and His Sources: The Heritage of the Collegio Romano in Galileo's Science* (Princeton: Princeton Univ. Press, 1984). [. . . .]

47   Christopher Clavius, *In Sphaeram Ioannis de Sacro Bosco Commentarius, Nunc Iterum ab Ipso Auctore Recognitus* (Venice, 1591), pp. 452–458. See also Nicholas Jardine, "The Forging of Modern Realism: Kepler and Clavius against the Sceptics," *Studies in History and Philosophy of Science* 10 (1979): 141–173.

concentric spheres, countered by arguing that, by the rules of the syllogism, a true conclusion could be deduced from false premises; one could, in short, posit any set of eccentrics and epicycle – even if they were not true causes – so long as they saved the phenomena. Clavius answered that this objection *ex falsa verum* could just as well be applied to physics and to all the other disciplines, including theology. An unacceptable skepticism would result. In the midst of this controversy, Clavius introduced the case of the Copernican theory, which successfully uses epicycles and eccentrics to save the phenomena. Are we thereby left with another skeptical dilemma, unable now to choose between Copernicus and Ptolemy? In such a case of (what we would perhaps call) inter-theoretic conflict, Clavius argues, one respects the traditional disciplinary hierarchy and turns to natural philosophy and theology for assistance in discovering true causes. The point is an interesting one because it reveals the limits to Clavius's assertion of the primacy of mathematics over the other disciplines. His deeper aim was to bring *concordance* between Ptolemaic astronomy and natural philosophy. Copernicus's mathematical harmonies were not alone sufficient to induce Clavius to seek out a fundamentally new kind of physics.

Clavius's authority was enormous. The astronomical text in which he stated his views was used throughout the Jesuit colleges. It is thus no surprise that when Jesuit theologians considered Diego de Zuñiga's Copernican reading of Job 9:6, they turned to Clavius for guidance. For example, when the Spanish Jesuit Juan de Pineda (1558–1637), philosopher and theologian, consultor to the Inquisition, and compiler of the Spanish Index arrived at the moot passage 9:6 in his *Commentary on Job*, he declared Zuñiga's interpretation to be "plainly false," as shown "elegantly with firm reasons from philosophy and astrology by Our Christopher Clavius." Scripturally, he followed Zuñiga's literalism but offered an alternative, literal reading, rendering *terrae commotio* not as change of place but as *terrae tremor* – a quivering or shaking of the earth, sign of God's power and displeasure with man.[48] In 1620, four years after the decree against Copernicus and Zuñiga, he published a commentary on Ecclesiastes in which the language was stronger: "Diego de Zuñiga, a man knowledgeable and distinguished in religion, was babbling idly when he wrote on Job 9:6.... [His views are] false and dangerous...."[49] Pineda's *Commentary on Job* was immediately influential. At the Collège de France the theologian Jean Lorin (1559–1634) wrote; "Our Clavius . . . and Pineda in Job chap. 9, demonstrate his [Zuñiga's] opinion to be false and rash. . . . This aforesaid opinion can be seen to be dangerous and repugnant to the Faith. . . ."[50] It is also noteworthy

---

48   *Commentariorum in Iob Libri Tredecim* (Cologne, 1600), p. 340; Wardęska Texts, p. 31.
49   *In Ecclesiasten Commentariorum Liber Unus* (Antwerp, 1620), pp. 111–118; Wardęska Texts, pp. 32a–h.
50   *In Acta Apostolorum Commentarii* (Lyon, 1606), p. 215; Wardęska Texts, p. 22.

that Cardinal Robert Bellarmine, who knew Lorin personally;[51] possessed a copy of Pineda's commentary in his library.[52] [. . .]

Copernican Theologies

In the atmosphere of literalism prevalent in the sixteenth century, the Copernicans had to address the question of the Bible's true sense when it uses the words *sun, earth,* or *moon*; they had to argue that the moral and symbolic meanings of these words were *detachable* from any literal reference to the physical world. But there are other passages in the Scriptures that do not mention the celestial bodies and yet were singled out as containing special meaning about God's relation to nature. Romans 1:20, appealed to by Christians in a tradition reaching back to Saint Augustine, provided a text capable of wide connotation, not least as a basis for legitimating alternatives to Aristotle's philosophy of nature: "For the invisible things of Him from the creation of the world are clearly seen, being understood by the things that are made. . . ." From this evolved an important metaphor, invoked by both Protestants and Catholics: nature is a book through which the invisible God reveals Himself sensibly to man. Now if the intention of the author of the Scriptures was also that of the author of nature, then there could be only one truth revealed – though in different forms of discourse. In what language, then, did God write the book of nature? And which disciplines would yield privileged access to its meaning? The sixteenth-century Copernicans were as diverse in their hermeneutic and disciplinary references as they were in their theological assumptions. A few abbreviated examples must suffice here.

The Puritan Thomas Digges (1546–1595?) was the earliest Englishman to offer a defense of the Copernican theory. This occurs in the form of a very brief treatise appended to an almanac written by his father, Leonard Digges.[53] Accompanying Digges's account is a diagram of the universe portraying the heliocentric system surrounded by the orb of fixed stars, described by Digges

---

51   See James Brodrick, S.J., *The Life and Work of Blessed Robert Francis Cardinal Bellarmine, S.J., 1542–1621,* 2 vols. (London: Burns & Oates Ltd., 1928), 1:212.
52   This may be inferred from the "Index of Explicators of Holy Scripture" appended to the *Opera Omnia* of Bellarmine, ed. Justinus Fevre, 12 vols. (Paris: Vivès, 1870–1874; reprint, Frankfurt a.M.: Minerva, 1965), 12:478.
53   Leonard Digges, Gentleman, *A Prognostication euerlastinge of righte good effecte . . . Lately corrected and augmented by Thomas Digges his sonne* (London: Thomas Marsh, 1576). The supplement is entitled: *A Perfit Description of the Caelestiall Orbes according to the most aunciente doctrine of the Pythagoreans, latelye reuiued by Copernicus and by Geometricall Demonstrations approued,* ed. with commentary by Francis R. Johnson and Sanford V. Larkey, "Thomas Digges, the Copernican System, and the Idea of the Infinity of the Universe in 1576," *The Huntington Library Bulletin* 5 (1934): 69–117.

as infinitely extended in all dimensions. [. . .] If we examine the picture carefully, however, we see that it is not merely a cosmological representation but a
hieroglyph with a soteriological message. The orb carrying the earth is labeled
the "Globe of Mortality," that is, man in a state of sin after the Fall. The fixed
sun lights up the darkness of man's understanding, as Digges writes, "sphaer-
ically dispearsing his glorious beames of light through al this sacred
Coelestiall Temple." And the stars are the realm of the saved: "the very court
of coelestiall angelles devoid of greefe and replenished with perfite endlesse
love the habitacle for the elect." Digges's account thus portrays the world both
as a harmonious order displayed in the Copernican arrangement and as the
symbolic stage of a great Puritan drama. [. . .]

Galileo: Progressive Catholic Reformer

It is illuminating to regard Galileo, if only briefly, against this somewhat
complex career of the Copernican theory in the sixteenth century. To begin
with: he was, at first, an *academic* Copernican.[54] This in itself was fairly
unusual; apart from Zuñiga and Rheticus the only really comparable example
was Kepler's teacher, Michael Maestlin (1550–1631), at Tübingen; and
Maestlin, like the early Galileo, was fairly cautious about polemicizing on
behalf of Copernicus.[55] Second, we now know that Galileo's astronomical
lectures at Pisa were largely paraphrases of Clavius's *Commentary on the
Sphere.*[56] Thus, Galileo's early exposure to the Copernican problem would
have been conditioned by Clavius and his arguments against philosophers at
the Jesuit Collegio Romano who insisted that astronomy could never achieve

---

54   The extent of our *direct* knowledge of Galileo's Copernican views is contained in a well-
known letter to Kepler, 4 Aug. 1597: "I have for many years past considered the view of
Copernicus, and from it I have been able to discover the causes of many natural phenomena which
without doubt cannot be explained by the traditional hypotheses. I have drawn up a list of many
reasons and refutations of contrary arguments which, however, I thus far do not dare to make
public, being frightened by the fate of our teacher Copernicus, who, having gained immortal fame
in the eyes of a few, has been ridiculed and hissed off the stage by innumerable others (for so great
is the number of fools). I should indeed venture to disclose my thoughts if there were more men
like you; since there are none, I shall desist from such a venture" (*Le opere di Galileo Galilei*, ed.
Antonio Favaro, 20 vols. [Florence: G. Barbèra, 1899–1909], 10:68). [. . .]
55   See Robert S. Westman, "Michael Mästlin's Adoption of the Copernican Theory," in *Colloquia
Copernicana*, vol. 4, Studia Copernicana, vol. 13 (Wrocław: Ossolineum, 1975), pp. 51–61, and
"Three Responses," pp. 329–337.
56   See A. C. Crombie, "Sources of Galileo's Early Natural Philosophy," in *Reason, Experiment and
Mysticism in the Scientific Revolution*, ed. M. L. Righini Bonelli and William R. Shea (New York:
Science History Publications, 1975), pp. 157–175; William A. Wallace, *Prelude to Galileo: Essays
on Medieval and Sixteenth-Century Sources of Galileo's Thought* (Dordrecht: Reidel, 1981), pp.
192–252. [. . .]

certitude.[57] Third, and most significantly, in his defense of the Copernican system Galileo committed himself to a strict notion of proof in science, according to which true conclusions must be deduced necessarily from true premises, which are themselves self-evident.[58] No other Copernican had locked himself into such a tight position. Fourth, Galileo's advocacy of demonstrative proof in the Copernican matter was deeply affected by both scientific and political developments. The sudden availability of the telescope in 1609[59] radically changed what it was now possible to do in the domain of celestial natural philosophy. Among other things, it provided a conclusive argument against the Ptolemaic theory by proving the existence of phases of Venus. This success, which even some Jesuits at the Collegio Romano were willing to acknowledge, boosted Galileo's confidence that a grand demonstration could eventually be found that would establish the Copernican system conclusively against other alternatives.

Galileo used his success with the telescope to negotiate a return to Florence – to leave his academic position in Venice, where he was now very secure, well paid, and relatively free from church interference – and to take up a position at the court of the grand duke of Tuscany.[60] He made his move in 1610, against the better judgment of his Venetian friends,[61] and was awarded the title "Chief Mathematician and Philosopher" at his own request.[62] He also received an appointment from the grand duke as chief professor of mathematics at the University of Pisa *without obligation to teach or reside there*.[63] He was now free to break with fixed forms of academic disputation and to express himself in the vernacular Italian of his native Tuscany.[64] No other Copernican

57   The case for Clavius's importance in providing the context for Galileo's earliest methodological convictions has been made eloquently by William A. Wallace (*Prelude*, pp. 137–138, 231–233; *Galileo and His Sources*, pp. 126–148).

58   Historians continue to debate the ways in which Galileo's various proof structures are to be characterized and the contrast between his ideal of science and the way in which he practiced science. [. . .]

59   See Albert Van Helden, *The Invention of the Telescope*, Transactions of the American Philosophical Society, vol. 67, no. 4 (Philadelphia: American Philosophical Society, 1977).

60   The telescope provided Galileo with the institutional leverage that he would need to leave the university and to embark on a series of topics that would have met with resistance from the entrenched disciplinary hierarchy of the university. See Galileo to Belisario Vinta, quoted and translated in Wallace, *Prelude*, p. 139.

61   Galileo's Venetian friend Francesco Sagredo, immortalized in the *Dialogue*, viewed Galileo's move as an error; see his argument, translated in Stillman Drake, *Discoveries and Opinions of Galileo* (Garden City, NY: Doubleday Anchor, 1957), pp. 67 ff.

62   "As to the title of my position," Galileo wrote to the grand duke, "I desire that in addition to the title of 'mathematician' His Highness will annex that of 'philosopher'; for I may claim to have studied more years in philosophy than months in pure mathematics" (translated in Drake, *Discoveries and Opinions*, p. 64).

63   Ibid., pp. 61–62.

64   "The Tuscan language," wrote Galileo, "is entirely adequate to treat and to explain the concepts of all branches of knowledge" (Galileo, *Opere* 5:189–190). [. . .]

except Kepler had hitherto engineered a position with such powerful disciplinary freedom and leverage.

The move to the Florentine court did not, however, provide Galileo with all the immunity from attack that he had hoped to achieve. Although he had received a warm reception from the Jesuits for his telescopic discoveries, he encountered hostility in Florence from two quarters: certain philosophers at Pisa, most notably Lodovico delle Colombe (1565 – ca. 1615) and a Dominican preacher, Tommaso Caccini (1574–1648), who was in possession of Tolosani's long-buried, anti-Copernican treatise.[65] Delle Colombe, arguing from the traditional view of the disciplinary superiority of philosophy and theology, claimed that mathematics offered no certitude about what moves and what does not, because it abstracts from matter, whereas philosophy is concerned with essences. In short, one should not assume, as Copernicus did, that the earth moves in order to find out what consequences follow; rather, one asks first whether it is self-evident that the earth the is physically suitable for motion. And with regard to Scripture, delle Colombe declared: "All theologians without a single exception say that when Scripture can be understood according to the literal sense, it must never be interpreted in any other way."[66] Caccini's attack was even less temperate. In December 1614 he delivered a sermon denouncing mathematics as a diabolical art, mathematicians as violators of the Christian religion and enemies of the state, and Galileo as the propagator of cosmological absurdities. He cleverly cited a passage from Acts 1:11: "Ye Men of Galilee, why stand ye gazing up into heaven?" thus punning on Galileo's name.[67] It was a popular sermon, not a learned disputation like Tolosani's critique of 1546 or Clavius's commentary of 1581. Galileo's disciple, the Benedictine Benedetto Castelli, who held the lower mathematics position at Pisa, well described the quality of the opposition when he wrote of "those pickpockets and highwaymen who waylay mathematicians."[68]

The Copernican situation in Italy during this period, as we now see, does not lend itself to facile dichotomies. The most powerful order, the Society of Jesus, was opposed to Copernicanism but not unalterably closed to the novelties revealed by the telescope. Indeed, there were further important divisions within the order itself: some Jesuits were far more willing to criticize Aristotle than is usually recognized.[69] Two nonacademic, reform-minded Dominicans

65   See notes 36 and 37; an annotation by Caccini appears on fol. 3r of the Tolosani MS.

66   Galileo, *Opere* 3:255, 290.

67   See Antonio Ricci-Riccardi, *Galileo Galilei e Fra Tommaso Caccini* (Florence: Successori Le Monnier, 1902), pp. 66–67; Jerome J. Langford, *Galileo, Science and the Church* (Ann Arbor, Mich.: Univ. of Michigan Press, 1966), p. 55; Drake, *Discoveries and Opinions*, pp. 153–154.

68   Galileo *Opere* 12:123. Castelli's comment suggests again that the opposition was not confined to academics and that there was an element of Florentine popular street sentiment. [. . .]

69   See Ugo Baldini, "*Additamenta Galileiana*, I: Galileo, La nuova astronomia e la critica all'aristotelismo nel dialogo epistolare tra Giuseppe Biancini e i Revisori Romani della Compagnia di Gesu," *Annali dell'Istituto e Museo di Storia della Scienza di Firenze* 9 (1984): 13–43.

from the socially tumultuous region of Naples, Bruno and Tommaso Campanella, had given vigorous support, respectively, to Copernicus and to Galileo – although both had come to bad ends for their political views.[70] Castelli the Benedictine and, later, Bonaventura Cavalieri, a Jesuate, were faithful Galilean disciples. Zuñiga at Salamanca had been an Augustinian. Now, in January 1615, came unexpected support for the Copernican position from a member of the reformed Carmelite order in Naples, Paolo Foscarini (1580–1616). Foscarini was well read; and, indeed, by 1615 there was more to read than there had been when Zuñiga wrote his commentary on Job in the early 1580s: Galileo's *Sidereal Messenger*, Kepler's *New Astronomy*, Clavius's last edition of the *Sphere*, suggesting the need for a new cosmology, and Copernicus's *De Revolutionibus*. Foscarini knew and cited them all. His *Letter Concerning the Opinion of Copernicus and the Pythagoreans about the Mobility of the Earth* is an ambitious garnering of all biblical texts bearing on the Copernican theory with the object of showing that they can be reconciled with heliocentrism. Building primarily on the evidence of Galileo's telescopic observations of Venus, he rejects Ptolemy's theory, maintains that all physical propositions in the Bible are absolutely accommodated to popular discourse and concludes that the Copernican "opinion" is "not improbable."[71] Foscarini treats the problem of cosmological choice as though it were a moral problem, falling therefore into the domain of the contingent or that which can be other than it is. Under such conditions, what we can have is *opinionative* knowledge, the best expression of truth at that time – rather than *demonstrative knowledge*, that which must follow all the time from true premises.[72]

Thus, Galileo was not alone. Though not a cleric, he belongs by family affinity within a small movement of church progressives, not all Copernicans, who sought reform of traditional positions on natural knowledge and more liberal rules of scriptural translation and interpretation. They included a considerable range: from the Salamancan humanists Luis de León and Diego de Zuñiga to the Neapolitan probabiliorist Foscarini; from the moderate academic mathematicism of Clavius at the Collegio Romano[73] to the heterodox

---

70   On the political situation in Naples at the end of the sixteenth century see Rosario Villari, "Naples: The Insurrection in Naples in 1585," in *The Late Italian Renaissance, 1525–1630*, ed. Eric Cochrane (New York: Macmillan, 1970), pp. 305–330; cf. Carolyn Merchant, *The Death of Nature: Women, Ecology and the Scientific Revolution* (San Francisco: Harper & Row, 1980), pp. 115–117.

71   Paolo Foscarini, in Wardęska Texts, pp. 10a–z, esp. 10e and 10z.

72   For an excellent discussion of the concept of probability and moral deliberation in Thomas Aquinas, whose influence on this issue was considerable, see Edmund F. Byrne, *Probability and Opinion: A Study in the Medieval Presuppositions of Post-Medieval Theories of Probability* (The Hague: Martinus Nijhoff, 1968), pp. 213–227.

73   As William Wallace has emphasized, expanding on the work of R. G. Villoslada, there were close connections between the Spanish Jesuits and Dominicans, especially at Salamanca and the Collegio Romano. Clavius himself had studied under Pedro Nuñez at the Portuguese university in

Neapolitan metaphysics of friars Bruno and Campanella; and finally, Galileo's numerous, mathematically talented academic disciples in the minor orders. Against this group was the conservative heritage of Trent – represented most formidably by the greatest Jesuit controversialist of the period, Cardinal Robert Bellarmine (1542–1621) – although the young Bellarmine had entertained quite remarkably un-Aristotelian and even un-Ptolemaic notions in his Louvain lectures of 1570–1572.[74] Bellarmine was an old friend and colleague of Clavius, author of the preface to the Clementine Vulgate of 1592 (the very symbol of Tridentine authority), a member of the commission that tried and convicted Bruno in 1599,[75] and head of the Collegio Romano in 1611, when it honored Galileo for his telescopic discoveries. It was the same Bellarmine to whom Foscarini sent his little book in 1615. Yet, in his well-known reply to Foscarini, Bellarmine thundered with the weight of tradition: "The Council [of Trent] would prohibit expounding the Bible contrary to the common agreement of the Fathers." The Fathers *do* agree, according to Bellarmine, on a literal reading of the standard passages concerning cosmology. It is, therefore, a matter of faith because of the unanimity of those who have spoken. Mathematicians should therefore restrict themselves to speaking "hypothetically and not absolutely," for to speak absolutely would injure the faith and irritate all the theologians and Scholastic philosophers. Only if there were a strict demonstration would one be permitted to accommodate Scripture absolutely to popular discourse – and Bellarmine was convinced that no such demonstration existed.[76]

Galileo obtained a copy of Bellarmine's letter and immediately recognized the rejection of Foscarini's probabiliorism.[77] Equally significant, he accepted the conditions of strict demonstration that Bellarmine laid down and which, in any case, he had accepted long before under the influence of his associations with the Collegio Romano. "Our opinion," he wrote in notes to the Bellarmine letter, "is that the Scriptures accord perfectly with demonstrated

Coimbra, and there existed a general tendency for the Roman Jesuits to imitate Parisian and Salamancan academic styles (Wallace, *Prelude*, pp. 229, 241).

74 Ugo Baldini and George V. Coyne, S.J., "The Louvain Lectures (Lectiones Lovanienses) of Bellarmine and the Autograph Copy of His 1616 Declaration to Galileo: Texts in the Original Latin (Italian) with English Translation, Introduction, Commentary and Notes," *Vatican Observatory Publications*, Special Series: *Studi Galileiani*, 1 (1984): 3–48.

75 Xavier Le Bachelet has established that Bellarmine actually played a lesser role in the trial of Bruno ("Bellarmin et Giordano Bruno," *Gregorianum* 4 [1923]: 193–210).

76 Galileo, *Opere* 12:171–172; Drake, *Discoveries and Opinions*, pp. 162–164.

77 We know that Galileo had received a copy of Foscarini's book from his friend Federico Cesi by mid-March 1615 and that the book had been read approvingly by his friends in Rome (pp. 162–163) and in Pisa (p. 165). Much of what we know about Galileo's immediate response to Bellarmine is based upon a critical undated document, apparently composed sometime between March and June 1615 ("Considerazione circa l'opinione Copernicana"), translated in Drake, *Discoveries and Opinions*, pp. 167–170.

physical truth. But let those theologians who are not astronomers guard against rendering the Scriptures false by trying to interpret against it propositions which may be true and might be proved so."[78] This brief remark contains much of the gist of Galileo's position developed in his *Letter to the Grand Duchess Christina* (composed in June 1615, not published until 1635). Although Galileo readily appealed to the book of nature as warrant for his mathematical mode of philosophizing about nature, it is interesting to observe that, unlike Bruno and Kepler, he would not try to create a systematic theological justification for the Copernican arrangement. Instead, he chose to fight the battle at the boundaries of natural philosophy, mathematics, and biblical theology. But the *Letter*, though filled with the *rhetoric* of necessary demonstration and sense experience, failed to specify the true and necessary cause of the earth's motion – even though Galileo had at that time an argument about the tides that he believed to be conclusive.[79] Rather, the principal strategy of the *Letter* was to argue against the disciplinary hegemony of the conservative theologians and philosophers.[80] By this tactic Galileo hoped to convict the conservatives of two charges: first, theology is not "queen of the sciences" by virtue of providing the principles on which the less sublime disciplines are founded. It gains its authority through the dignity of its subject matter, which is eternal and sublime, and the special means by which it communicates to men (revelation). Second, the decree of the fourth session of Trent had been improperly interpreted in the matter of the unanimous consent of the Fathers and thus failed the test of demonstrative truth *in theology*. The Fathers, says Galileo, never debated the question of the motion of the earth. It was not even controversial. Their statements about the stability of the earth were all accommodated to the language of the people. "Hence it is not sufficient to say that because all the Fathers admitted the stability of the earth, this is a matter of faith; one would have to prove also that they had condemned the contrary opinion."[81]

---

78   Ibid., p. 168.
79   In the *Letter to Christina* Galileo employs various locutions: "necessarie dimostrazioni," "certezza di alcune conclusioni naturali," "le causi de quali forse inaltro modo non possono assegnare," etc. (*Opere* 5:311, 316–317). But nowhere does he refer to arguments contained in his *Dialogue on the Tides*, a work completed by January 1616 and later to become Day Four of his *Dialogue Concerning the Two Chief World Systems* (1632). On the tidal arguments see esp. William Shea, *Galileo's Intellectual Revolution: Middle Period, 1610–1632* (New York: Science History Publications, 1972), pp. 172–189.
80   As Galileo puts the point: "Why, this would be as if an absolute despot [i.e., the superior science], being neither a physician nor an architect but knowing himself free to command, should undertake to administer medicine and erect buildings according to his whim – at grave peril of his poor patients' lives, and the speedy collapse of his edifices" (Drake, *Discoveries and Opinions*, p. 193).
81   Ibid., p. 202. See also the very fine analysis of Olaf Pedersen that appeared as this article underwent revisions: "Galileo and the Council of Trent: The Galileo Affair Revisited," *Journal for the History of Astronomy* 14 (1983): 1–29, esp. 16–24.

At this point in his discussion Galileo introduced the testimony of Diego de Zuñiga. "Some theologians," Galileo wrote, "have now begun to consider it [the earth's motion] and they are seen not to deem it erroneous."[82] In so doing, he thereby gave implicit approval to the partial-accommodationist line taken by Zuñiga and identified himself with the temper of liberal Salamancan humanism. He could have no way of knowing that a copy of Pineda's critique of Zuñiga's book was in the library of Cardinal Bellarmine. It is significant that his citation of Zuñiga is then followed immediately by a veiled reference to delle Colombe's very conservative interpretation of the Council's position that physical conclusions, when there is a consensus of the Fathers, are a matter of faith. "I think this may be an arbitrary simplification of various council decrees by certain people to favor their own opinion,"[83] wrote Galileo. The motion of the earth, he was trying to say, is not a matter of faith and morals.

The *Letter* ends with a subtle exegesis of Joshua 10:12–13. The logic of Galileo's position here cleverly seeks to shift the exegetical ground to the *astronomical* meaning of the passage, thereby exposing the theologian's astronomical incompetence even in the area of their own subject matter and presaging the structure of his arguments in the *Dialogue Concerning the Two Chief World Systems* (1632). *Even if* physical propositions were a matter of faith, which they are not, astronomically informed theologians could still ask whether Joshua's request for more daylight might be better accomplished on the Copernican system than on the Ptolemaic. More cautious than Zuñiga, Galileo was nevertheless close to the Spaniard in flirting with the construction of the dangerous geomotive interpretation of Joshua.

Conclusion

The official Catholic response to Copernicus's theory on 5 March 1616 masks the very complex history that we have constructed – the diversity of Copernican discourses, the variety of disciplinary and exegetical strategies employed by Protestants and Catholics, and finally the struggle *within the church itself* between reformers and traditionalists.[84] The decree published by the Congregation of the Index ordering that Copernicus's *De Revolutionibus* and Zuñiga's *Commentary on Job* be "suspended until corrected" represented a local victory for the conservative Tridentine faction of the church, for the maintenance of traditional hierarchical authority in the universities and

---

82   Galileo continues: "Thus in the *Commentaries on Job* of Didacus a Stuñica, where the author comments upon the words *Who moveth the earth from its place* . . . , he discourses at length upon the Copernican opinion and concludes that the mobility of the earth is not contrary to Scripture" (Drake, *Discoveries and Opinions*, p. 203).

83   Ibid.

84   Galileo, *Opere* 19:322–323.

within the Jesuit Order. What was the immediate effect of this decree? Catholics could buy and read the books, but they were to know that the doctrines contained therein were false, and they were instructed on which lines to expurgate.[85] The decree was hard to enforce. Of more than five hundred copies of *De Revolutionibus* still extant in the world today, only about 8 percent have been censored by their seventeenth-century owners.[86] The offensive passages in the copy of Zuñiga's extremely rare book in the library of the University of Salamanca have been heavily crossed out. The British Library copy has thick paper pasted over the dangerous sections – so nicely done, in fact, that the book automatically falls open to that section of the commentary! We are uninformed as to other extant copies. Paolo Foscarini's little treatise was "altogether prohibited and condemned." His work was perceived as more threatening because its extensive accommodations of Scripture to the Copernican theory were not as easily handled with scissors, paste, and ink. The name Galileo Galilei appears nowhere in the decree.

85   The actual instructions for correcting copies of *De Revolutionibus* were not published until 15 May 1620 (ibid., 19:400–401).
86   See Owen Gingerich, "The Censorship of Copernicus' *De Revolutionibus*," *Annali dell'Istituto e Museo di Storia della Scienza di Firenza* 7 (1981): 45–61.

3

The Experimental Philosophy and Its Institutions

Pump and Circumstance: Robert Boyle's Literary Technology

Steven Shapin

Originally appeared as "Pump and Circumstance: Robert Boyle's Literary Technology," in *Social Studies of Science*, edited by H. M. Collins, T. J. Pinch, and Steven Shapin Vol. 14 (London, Beverly Hills, and New Delhi: Sage, 1984): 481–520.

The text and notes in this chapter have been abridged. For complete footnotes the original publication should be referred to.

Editor's Introduction

As we have seen, Hooykaas emphasized the importance of facts in what he termed the rise of modern science. But what is a fact and where does it come from? These were particularly acute problems in the seventeenth century because of the invention of new instruments that offered new ways of practicing natural philosophy. Certainly there had been instruments before the seventeenth century, but these can be termed mathematical instruments, that is, they measured phenomena that were readily accessible to the unaided observer. What was new, however, were instruments which produced phenomena to which somebody without that instrument had no access. The most celebrated of these instruments was the telescope. Galileo, the first to turn the telescope to the heavens, gained international celebrity with his discovery of the moons of Jupiter. In experimental natural philosophy, the most astonishing new instrument was the air-pump, which, its inventors claimed, could create a space devoid of air.

But there was a major problem confronting the astronomer looking through the telescope or the natural philosopher subjecting candles, bladders, bells, and birds to the air-pump – how do you convince others that what you say you see is actually there when you have the only example of that particular instrument in existence and they cannot see for themselves? While the telescope spread rapidly throughout Europe, when Robert Boyle began experimenting with the air-pump there was no other one like it in the world. Boyle was therefore quite aware of the problem of replication his readers would encounter; no one had the equipment to repeat the experiments Boyle described in his published work. How then was he to convince them of the truth of his claims? How would they know that his observations were reliable? In short, how could his claims become facts?

Steven Shapin, a sociologist of science who has spent much of his career investigating the issue of trust and credibility in science, takes up these questions here.[1] Shapin argues that facts are not discovered but created. Moreover, the production and communication of knowledge are not distinct; it is in the communication of claims that they become accepted as fact, that is, come to count as knowledge. In essence facts cannot exist without a community that holds them for credibility and trust can only occur within social relationships. The existence of facts and trust can also be promoted by particular social institutions. It is no coincidence that the second half of the seventeenth century saw the founding of the first scientific societies, in particular the Royal Philosophical Society of London and the Royal Academy of Paris.

Boyle could publicly perform his experiments at the Royal Society where the Fellows of the Society could directly witness them and together establish matters of fact which all agreed upon. But to convince those people who were not physically present Boyle invented what Shapin terms material, literary, and social technologies to multiply the witnessing experience. That is, if the reader could not be an actual witness, by reading Boyle's book he could become a virtual witness.

---

1   For a more extended treatment see Shapin and Schaffer, *Leviathan and the Air-Pump: Hobbes, Boyle, and the Experimental Life* (Princeton: Princeton University Press, 1985; also Shapin, *A Social History of Truth: Civility and Science in Seventeenth-Century England* (Chicago: University of Chicago Press, 1994).

Pump and Circumstance: Robert Boyle's Literary Technology

Steven Shapin

The production of knowledge and the communication of knowledge are usually regarded as distinct activities. In this paper I shall argue to the contrary: speech about natural reality is a means of generating knowledge about reality, of securing assent to that knowledge, and of bounding domains of certain knowledge from areas of less certain standing. I shall attempt to display the conventional status of specific ways of speaking about nature and natural knowledge, and I shall examine the historical circumstances in which these ways of speaking were institutionalized. Although I shall be dealing with communication within a scientific community, there is a clear connection between this study and the analysis of scientific popularization. The popularization of science is usually understood as the extension of experience from the few to the many. I argue here that one of the major resources for generating and validating items of knowledge within the scientific community under study was this same extension of experience from the few to the many: the creation of a scientific public. The etymology of some of our key terms is apposite: if a *community* is a group sharing a common life, *communication* is a means of making things common.

The materials selected to address this issue come from episodes of unusual interest to the history, philosophy and sociology of science. Robert Boyle's experiments in pneumatics in the late 1650s and early 1660s represent a revolutionary moment in the career of scientific knowledge. In his *New Experiments Physico-Mechanical* (1660) and related texts of the early Restoration, Boyle not only produced new knowledge of the behaviour of air, he exhibited the proper experimental means by which legitimate knowledge was to be generated and evaluated. And he did so against the background of alternative programmes for the production of knowledge, the proponents of which subjected Boyle's recommended methods to explicit criticism. What was at issue in the controversies over Boyle's air-pump experiments during the 1660s was the question of how claims were to be authenticated as knowledge. What was to count as knowledge, or 'science'? How was this to be distinguished from other epistemological categories, such as 'belief' and 'opinion'? What degree of certainty could be expected of various intellectual enterprises and items of knowledge? And how could the appropriate grades of assurance and certainty be secured?[1]

---

1   R. Boyle, 'New Experiments Physico-Mechanical, touching the Spring of the Air . . .', in Boyle, *Works*, ed. T. Birch, 6 vols. (London, 1772), vol. I, 1–117. (All subsequent references to Boyle's writings are to this edition and will be cited as *RBW*.)

These were all practical matters. In the setting of early Restoration England there was no one solution to the problem of knowledge which commanded universal assent. The technology of producing knowledge had to be built, exemplified and defended against attack. The categories of knowledge and their generation that seem to us self-evident and unproblematic were neither self-evident nor unproblematic in the 1660s. The foundations of knowledge were not matters merely for philosophers' reflections; they had to be constructed and the propriety of their foundational status had to be argued. The difficulties that many historians evidently have in recognizing this work of construction arise from the very success of that work: to a very large extent we live in the conventional world of knowledge-production that Boyle and his colleagues amongst the experimental philosophers laboured to make safe, self-evident and solid.

Robert Boyle sought to secure universal assent by way of the experimental *matter of fact*. About such facts one could be highly certain; about other items of natural knowledge more circumspection was indicated. Boyle was, therefore, an important actor in the probabilist and fallibilist movement of seventeenth-century England. Before *circa* 1660, as Hacking and Shapiro have shown, the designations of 'knowledge' and 'science' were rigidly distinguished from 'opinion'.[2] Of the former one could expect the absolute certainty of *demonstration*, exemplified by logic and geometry. The goal of physical science had been to attain to this kind of certainty that compelled assent. By contrast, the English experimentalists of the mid-seventeenth century increasingly took the view that all that could be expected of physical knowledge was *probability*, thus breaking down the radical distinction between 'knowledge' and 'opinion'. Physical hypotheses were provisional and revisable; assent to them was not necessary, as it was to mathematical demonstration; and physical science was, to varying degrees, removed from the realm of the demonstrative.[3] The probabilistic conception of physical knowledge was not regarded as a regrettable retreat from more ambitious goals; it was celebrated by its proponents as a wise rejection of failed dogmatism. The quest for necessary and universal assent to physical propositions was seen as improper and impolitic.

---

2   I. Hacking, *The Emergence of Probability: A Philosophical Study of Early Ideas about Probability, Induction and Statistical Inference* (Cambridge: Cambridge University Press, 1975), esp. Chapters 3–5; B. J. Shapiro, *Probability and Certainty in Seventeenth-Century England: A Study of the Relationships between Natural Science, Religion, History, Law and Literature* (Princeton, NJ: Princeton University Press, 1983), esp. Chapter 2.

3   Newton's place in the development of a probabilist view of physical science is ambiguous. Certain of his critics thought that he aimed at the necessary assent which most English natural philosophers had agreed to eschew; see Z. Bechler, 'Newton's 1672 Optical Controversies: A Study in the Grammar of Scientific Dissent', in Y. Elkana (ed.), *The Interaction between Science and Philosophy* (Atlantic Highlands, NJ: Humanities Press, 1974), 115–42.

If universal assent was not to be expected of explanatory constructs in science, how, then, was proper science to be founded? Boyle and the experimentalists offered the *matter of fact*. The fact was the item of knowledge about which it was legitimate to be 'morally certain'. A crucial boundary was drawn around the domain of the factual, separating it from those items which might be otherwise and from which absolute and permanent certainty should not be expected. Nature was like a clock: man could be certain of its effects, of the hours shown by its hands; but the mechanism by which these effects were produced, the clock-work, might be various.[4]

It is in the understanding of how matters of fact were produced and how they came to command universal assent that historians have tended to succumb to the temptations of self-evidence.[5] It is the purpose of this paper to display the processes by which Boyle constructed experimental matters of fact and thereby produced the conditions in which assent could be mobilized.

The Mechanics of Fact-Making

Boyle proposed that matters of fact be generated by a multiplication of the witnessing experience. An experience, even of an experimental performance, that was witnessed by one man alone was not a matter of fact. If that witness could be extended to many, and in principle to all men, then the result could be constituted as a matter of fact. In this way, the matter of fact was at once an epistemological and a social category. The foundational category of the experimental philosophy, and of what counted as properly grounded knowledge generally, was an artefact of communication and of whatever social forms were deemed necessary to sustain and enhance communication. I argue that the establishment of matters of fact utilized three technologies: a *material technology* embedded in the construction and operation of the air-pump; a *literary technology* by means of which the phenomena produced by the pump were made known to those who were not direct witnesses; and a *social technology* which laid down the conventions natural philosophers should employ

4   The usual form in which Boyle phrased this was the statement that God might produce the same effects in nature through very different causes; therefore 'it is a very easy mistake for men to conclude that because an effect may be produced by such determinate causes, it must be so, or actually is so.' Boyle, 'Some Considerations touching the Usefulness of Experimental Natural Philosophy', *RBW*, vol. II, 1–201, at 45 (orig. publ, 1663). [. . .]

5   This is especially evident in historians' treatment (or lack thereof) of criticisms of seventeenth-century experimentalism by philosophers who denied both the central role of experimental procedures and the foundational status of the matter of fact. For example, insofar as Thomas Hobbes's criticisms of Boyle's experimental programme have been discussed, historians have preferred to conclude that he 'misunderstood' Boyle, or that he 'failed to appreciate' the power of experimental methods; see, among others, F. Brandt, *Thomas Hobbes' Mechanical Conception of Nature* (Copenhagen: Levin & Munksgaard, 1928), 377–78. [. . .]

in dealing with each other and considering knowledge-claims.[6] Given the concerns of this paper, I shall be devoting most attention to Boyle's literary technology: the expository means by which matters of fact were established and assent mobilized. Yet the impression should not be given that we are dealing with three distinct technologies: each embedded the others. For example, experimental practices employing the material technology of the air-pump crystallized particular forms of social organization; desired forms of social organization were dramatized in the exposition of experimental findings; the literary reporting of air-pump performances provided an experience that was said to be essential to the propagation of the material technology or even to be a valid substitute for direct witness. In studying Boyle's literary technology we are not, therefore, talking about something which is merely a 'report' of what was done elsewhere; we are dealing with a most important form of experience and the means for extending and validating experience.

The Material Technology of the Air-Pump

We start by noting the obvious: Boyle's matters of fact were *machine-made*. In his terminology, performances using the air-pump counted as 'unobvious' or 'elaborate' experiments, contrasted to either the 'simple' observation of nature or the 'obvious' experiments involved in reflecting upon common artefacts like the gardener's watering-pot.[7] The air-pump (or 'pneumatic engine') constructed for Boyle in 1659 (largely by Robert Hooke) was indeed an elaborate bit of scientific machinery [. . .].[8] It consisted of a glass 'receiver' of about 30-quarts volume, connected to a brass 'cylinder' ('3') within which plied a wooden piston or 'sucker' ('4'). The aim was to evacuate the receiver of atmospheric air and thus to achieve a working vacuum. This was done by manually operating a pair of valves: on the downstroke, valve 'S' (the stop-cock) was opened and valve 'R' was inserted; the sucker was then moved down by means of a rack-and-pinion device ('5' and '7'). On the upstroke, the stop-cock was

6   The use of the word 'technology' in reference to the 'software' of literary practices and social relations may appear jarring, but it is in fact etymologically justified, as Carl Mitcham nicely shows: C. Mitcham, 'Philosophy and the History of Technology', in G. Bugliarello and D. B. Doner (eds), *The History and Philosophy of Technology* (Urbana, Ill.: University of Illinois Press, 1979), 163–201, esp. 172 ff. [. . .]

7   See, for example, Boyle, 'An Examen of Mr. T. Hobbes his Dialogus Physicus de Natura Aëris . . .', in *RBW*, vol. I, 186–242, at 241 (orig. publ. 1662); Boyle, 'Animadversions upon Mr. Hobbes's Problemata de Vacuo', in *RBW*, vol, IV. 104–28, at 105 (orig. publ. 1674). [. . .]

8   Boyle described his pump in 'New Experiments', op. cit. note 1, 6–11. One of the best accounts of the original pump and subsequent designs is still G. Wilson, 'On the Early History of the Air-Pump in England', *Edinburgh New Philosophical Journal*, vol. 46 (1849), 330–54; see also R. G. Frank, Jr, *Harvey and the Oxford Physiologists: A Study of Scientific Ideas* (Berkeley, Calif.: University of California Press, 1980), 128–30.

closed, the valve 'R' removed, and a quantity of air drawn into the cylinder was expelled. This operation was repeated many times until the effort of moving the sucker became too great, at which point a working vacuum was deemed to have been attained. Great care had to be taken to ensure that the pump was sealed against leakage, for example at the juncture of receiver and cylinder and around the sides of the sucker. Experimental apparatus could be placed into the receiver through an aperture at the top of the receiver ('B–C'), for instance a barometer or simple Torricellian apparatus. The machine was then ready to produce matters of fact. Boyle used the pump to generate phenomena which he interpreted in terms of 'the spring of the air' (its elasticity) and the weight of the air (its pressure).

Boyle's air-pump was, as he said, an 'elaborate' device; it was also temperamental (difficult to operate properly) and very expensive: the air-pump was seventeenth-century 'Big Science'. To finance its construction on an individual basis it helped mightily to be a son of the Earl of Cork. Other natural philosophers, almost as well supplied with cash, shied away from the cost of having one built, and a major justification for founding scientific societies in the 1660s and afterwards was the collective financing of the instruments upon which the experimental philosophy was deemed to depend. Air-pumps were not widely distributed in the 1660s. They were scarce commodities: Boyle's original machine was quickly presented to the Royal Society of London; he had one or two re-designed instruments built for him by 1662, operating mainly in Oxford; Christiaan Huygens had one made in The Hague in 1661; there was one at the Montmor Academy in Paris; there was probably one at Christ's College, Cambridge by the mid-1660s, and Henry Power may have possessed one in Halifax from 1661. So far as can be found out, these were all the air-pumps that existed in the decade after their invention.[9]

Thus, air-pump technology posed a problem of access. If knowledge was to be produced using this technology, then the numbers of philosophers who could produce it were limited. Indeed, in Restoration England this restriction was one of the chief recommendations of 'elaborate' experimentation: knowledge could no longer legitimately be generated by alchemical 'secretists' and sectarian 'enthusiasts' who claimed individual and unmediated inspiration from God. Experimental knowledge was to be tempered by collective labour and disciplined by artificial devices. The very intricacy of machines like the air-pump allowed philosophers, it was said, to discern which cause, amongst the many possible, might be responsible for observed effects. This was something,

9   The only information we have concerning the cost of the Boyle pump indicates that a version of the *receiver* ran to £5: T. Birch, *The History of the Royal Society of London*, 4 vols. (London, 1756–1757), vol. II, 184. Given the expense of machining the actual pumping apparatus, an estimate of £25 for the entire engine might be conservative. Thus, an air-pump would have cost more than the annual salary of the Curator of the Royal Society, Robert Hooke, who was the London pump's chief operator. [. . .]

in Boyle's view, that the gardener's pot could not do.[10] However, access to the machine had to be opened up if knowledge-claims were not to be regarded as mere individual opinion and if the machine's matters of fact were not to be validated on the bare say-so of an individual's authority. How was this special sort of access to be achieved?

Witnessing Science

In Boyle's programme the capacity of experiments to yield matters of fact depended not only upon their actual performance but essentially upon the assurance of the relevant community that they had been so performed. He therefore made an important distinction between actual experiments and what are now termed 'thought experiments'.[11] If knowledge was to be empirically based, as Boyle and other English experimentalists insisted it should, then its experimental foundations had to be attested to by eye-witnesses. Many phenomena, and particularly those alleged by the alchemists, were difficult to credit; in which cases Boyle averred 'that they that have seen them can much more reasonably believe them, than they that have not.'[12] The problem with eye-witnessing as a criterion for assurance was one of discipline. How did one police the reports of witnesses so as to avoid radical individualism? Was one obliged to credit a report on the testimony of any witness whatever?

Boyle insisted that witnessing was to be a collective enterprise. In natural philosophy, as in criminal law, the reliability of testimony depended crucially upon its multiplicity:

> For, though the testimony of a single witness shall not suffice to prove the accused party guilty of murder; yet the testimony of two witnesses, though but of equal credit . . . shall ordinarily suffice to prove a man guilty; because it is thought reasonable to suppose, that, though each testimony single be but probable, yet a concurrence of such probabilities, (which ought in reason to be attributed to the truth of what they jointly tend to prove) may well amount to a moral certainty, i.e. such a certainty, as may warrant the judge to proceed to the sentence of death against the indicted party.[13]

10   Boyle, 'Examen of Hobbes', op. cit. note 7, 193. [. . .]
11   See, for example, Boyle, 'The Sceptical Chymist', in *RBW*, vol. I, 458–586, at 460 (orig. publ. 1661): here Boyle suggests that many 'experiments' reported by the alchemists 'questionless they never tried'. [. . .]
12   Boyle, 'Two Essays, Concerning the Unsuccessfulness of Experiments', in *RBW*, vol. I, 318–53, at 343 (orig. publ. 1661); Boyle, 'Sceptical Chymist', op. cit. note 11, 486. [. . .]
13   Boyle, 'Some Considerations about the Reconcileableness of Reason and Religion', in *RBW*, vol. IV, 151–91, at 182 (orig. publ. 1675); see also L. J. Daston, *The Reasonable Calculus: Classical Probability Theory, 1650–1840* (unpublished PhD dissertation, Harvard University, 1979), 90–91; on testimony: Hacking, op. cit. note 2, Chapter 3; on evidence in seventeenth-century English law,

And Thomas Sprat, defending the reliability of the Royal Society's judgements in matters of fact, inquired

> whether, seeing in all Countreys, that are govern'd by Laws, they expect no more, than the consent of two, or three witnesses, in matters of life, and estate; they will not think, they are fairly dealt withall, in what concerns their *Knowledg*, if they have the concurring Testimonies of *threescore or an hundred*.[14]

The thrust of the legal analogy should not be missed. It was not just that one was multiplying authority by multiplying witnesses (although this was part of the tactic); it was that right *action* could be taken, and seen to be taken, on the basis of these collective testimonies. The action concerned the positive giving of assent to matters of fact. The multiplication of witness was an indication that testimony referred to a true state of affairs in nature. Multiple witnessing was counted as an active, and not just a descriptive, licence. Does it not force the conclusion that such and such an action was done (a specific trial), and that subsequent action (offering assent) was warranted?

In experimental practice one way of securing the multiplication of witnesses was to perform experiments in a social space. The 'laboratory' was contrasted to the alchemist's closet precisely in that the former was said to be a public and the latter a private space. The early air-pump trials were routinely performed in the Royal Society's ordinary public rooms, the machine being brought there specially for the occasion.[15] In reporting upon his experimental performances Boyle commonly specified that they were 'many of them tried in the presence of ingenious men', or that he made them 'in the presence of an illustrious assembly of virtuosi (who were spectators of the experiment).'[16] Boyle's collaborator Robert Hooke worked to codify the Society's procedures for the standard recording of experiments: the register was 'to be sign'd by a certain Number of the Persons present, who have been present, and Witnesses of all the said Proceedings, who, by Sub-scribing their Names, will prove undoubted Testimony . . .'[17] And Sprat described the role of the 'Assembly' in

see Shapiro, op. cit. note 2, Chapter 5; S. Schaffer, 'Making Certain (essay review of Shapiro), *Social Studies of Science*, vol. 14 (1984), 137–52, esp. 146–47 (for the legal analogy of scientific witnessing).

14   T. Sprat, *History of the Royal Society* (London, 1667), 100.

15   One of the ways by which Hobbes attacked the experimental programme was to insinuate that the Royal Society was *not* a public place: not everyone could come to witness experimental displays; see T. Hobbes, 'Dialogus physicus de natura aeris . . .', in Hobbes, *Opera philosophica*, ed. Sir William Molesworth, 5 vols. (London, 1839–45), vol. IV, 233–96, at 240 (orig. publ. 1661). [. . .]

16   Boyle, 'New Experiments', op. cit. note 1, 1; Boyle, 'The History of Fluidity and Firmness', in *RBW*, vol. I, 377–442, at 410 (orig. publ. 1661); Boyle 'Defence against Linus', op. cit. note 11, 173.

17   R. Hooke, *Philosophical Experiments and Observations* (London, 1726), 27–28.

'resolv[ing] upon the matter of *Fact'* by collectively correcting individual idio-
syncracies of observation and judgement.[18] In reporting experiments that
were particularly crucial or problematic, Boyle named his witnesses and stipulated their qualifications. Thus, the experiment of the original air-pump trials
that was 'the principal fruit I promised myself from our engine' was conducted
in the presence of 'those excellent and deservedly famous Mathematic Professors, Dr *Wallis*, Dr *Ward*, and Mr *Wren* . . . , whom I name, both as justly count-
ing it an honour to be known to them, and as being glad of such judicious and
illustrious witnesses of our experiment . . .' Another important experiment
was attested to by Wallis 'who will be allowed to be a very competent judge in
these matters.' And in his censure of the alchemists Boyle generally warned
natural philosophers not 'to believe chymical experiments . . . unless he, that
delivers that, mentions his doing it upon his own particular knowledge, or
upon the relation of some credible person, avowing it upon his own experience.' Alchemists were recommended to name the putative author of these
experiments 'upon whose credit they relate' them.[19] The credibility of witnesses followed the taken-for-granted conventions of that setting for assessing
individuals' reliability and trustworthiness: Oxford professors were accounted
more reliable witnesses than Oxfordshire peasants. The natural philosopher
had no option but to rely for a substantial part of his knowledge on the testimony of witnesses; and, in assessing that testimony, he (no less than judge or
jury) had to determine their credibility. This necessarily involved their moral
constitution as well as their knowledgeableness, 'for the two grand requisites,
of a witness [are] the knowledge he has of the things he delivers, and his faith-
fulness in truly delivering what he knows.' Thus, the giving of witness in experimental philosophy transitted the social and moral accounting systems of
Restoration England.[20]

Another important way of multiplying witnesses to experimentally produced phenomena was to facilitate their replication. Experimental protocols
could be reported in such a way as to enable readers of the reports to perform
the experiments for themselves, thus ensuring distant but direct witnesses.
Boyle elected to publish several of his experimental series in the form of letters
to other experimentalists or potential experimentalists. The *New Experiments*
of 1660 was written as a letter to his nephew Lord Dungarvan; the various
tracts of the *Certain Physiological Essays* of 1661 were written to another
nephew Richard Jones; the *History of Colours* of 1664 was originally written
to an unspecified friend. The purpose of this form of communication was

18   Sprat, op. cit. note 14, 98–99; see also Shapiro, op. cit. note 2, 21–22.
19   Boyle, 'New Experiments', op. cit. note 1, 33–34; Boyle, 'A Discovery of the Admirable
Rarefaction of Air . . .', in *RBW*, vol. III, 496–500, at 498 (orig. publ. 1671); Boyle, 'Sceptical
Chymist', op. cit. note II, 460.
20   Boyle, 'The Christian Virtuoso', in *RBW*, vol. V, 508–40, at 529 (orig. publ. 1690); see also
Shapiro, op. cit. note 2, Chapter 5 (esp. 179). [. . .]

explicitly to proselytize. The *New Experiments* was published so 'that the person I addressed them to might, without mistake, and with as little trouble as possible, be able to repeat such unusual experiments . . .'. The *History of Colours* was designed 'not barely to relate [the experiments], but . . . to teach a young gentleman to make them.'[21] Boyle wished to encourage young gentlemen to 'addict' themselves to experimental pursuits and, thereby, to multiply both experimental philosophers and experimental facts.

Replication, however, rarely succeeded, as Boyle himself recognized. When he came to prepare the *Continuation of New Experiments* seven years after the original air-pump trials, Boyle admitted that, despite his care in communicating details of the engine and of his procedures, there had been few successful replications:

> . . . in five or six years I could hear but of one or two engines that were brought to be fit to work, and of but one or two new experiments that had been added by the ingenious owners of them . . .[22]

This situation had not notably changed by the mid-1670s. In the seven or eight years after the *Continuation*, Boyle said that he heard 'of very few experiments made, either in the engine I used, or in any other made after the model thereof.' By this time a note of despair began to appear in Boyle's statements concerning the replication of his air-pump experiments. He

> was more willing to set down divers things with their minute circumstances; because I was of opinion, that probably many of these experiments would be never either re-examined by others, or re-iterated by myself. For though they may be easily read . . . yet he, that shall really go about to repeat them, will find it no easy task.[23]

The Literary Technology of Virtual Witnessing

The third way by which witnesses could be multiplied is far more important than the performance of experiments before direct witnesses or the facilitating of actual replication: it is what I shall call 'virtual witnessing'. The technology of virtual witnessing involves the production in a reader's mind of such an

21   M. Boas [Hall], *Robert Boyle and Seventeenth-Century Chemistry* (Cambridge: Cambridge University Press, 1958), 40–41; Boyle, 'New Experiments', op. cit. note 1, 2; Boyle, 'The Experimental History of Colours', in *RBW*, vol, I, 662–778, at 633 (orig. publ. 1663). [. . .]

22   Boyle 'A Continuation of New Experiments Physico-Mechanical, touching the Spring and Weight of the Air', in *RBW*, vol. III, 175–276, at 176. This was written in 1668 and printed a year later. [. . .]

23   Boyle, 'A Continuation of New Experiments, Physico-Mechanical . . . The Second Part', in *RBW*, vol. IV, 505–93, at 505, 507 (orig. pub. 1680).

image of an experimental scene as obviates the necessity for either its direct witness or its replication. Through virtual witnessing the multiplication of witnesses could be in principle unlimited. It was therefore the most powerful technology for constituting matters of fact. The validation of experiments, and the crediting of their outcomes as matters of fact, necessarily entailed their realization in the laboratory of the mind and the mind's eye. What was required was a technology of trust and assurance that the things had been done and done in the way claimed.

The technology of virtual witnessing was not different in kind to that used to facilitate actual replication. One could deploy the same linguistic resources in order to encourage the physical replication of experiments or to trigger in the reader's mind a naturalistic image of the experimental scene. Of course, actual replication was to be preferred, for this eliminated reliance upon testimony altogether. Yet, because of natural and legitimate suspicion amongst those who were neither direct witnesses nor replicators, a greater degree of assurance was required to produce assent in virtual witnesses. Boyle's literary technology was crafted to secure this assent.

Prolixity and Iconography

In order to understand how Boyle deployed his literary technology of virtual witnessing we have to reorientate some of our common ideas about the status of the scientific *text*. We usually think of an experimental report as a narration of some prior visual experience: it points to sensory experience that lies behind the text. This is correct. However, we should also appreciate that the text itself constitutes a visual source. It is my task here to see how Boyle's texts were constructed so as to provide a source of virtual witness that was agreed to be reliable. The best way to fasten upon the notion of the text as this kind of source might be to start by looking at some of the pictures that Boyle provided alongside his prose.

[Boyle appended an engraving of his original air-pump to the *New Experiments*.] Producing these kinds of images was an expensive business in the mid-seventeenth century and natural philosophers used them sparingly. [This engraving] is not a schematized line-drawing but an attempt at detailed naturalistic representation, complete with the conventions of shadowing and cut-away sections of parts. This is not a picture of the 'idea' of an air-pump but of a particular existing air-pump.[24] The same applies to Boyle's pictorial representations of his particular pneumatic experiments: in one, we are shown a

24   This practice can be contrasted with the iconography of the anti-experimentalist Hobbes whose natural philosophy texts included only a few images of experimental systems, and these very simple and highly stylized. In giving his account of the air-pump and how it worked, Hobbes deliberately scorned the use of pictures; see Hobbes, op. cit. note 15, 235, 242. [. . .]

mouse lying dead in the receiver; in another, images of the experimenters. Boyle devoted great attention to the manufacture of these engravings, sometimes consulting directly with artist and engraver, sometimes by way of Hooke.[25] Their role was to be a supplement to the imaginative witness provided by the words in the text. In the *Continuation* Boyle expanded upon the relationships between the two sorts of exposition. He told his readers that 'they who either were versed in such kind of studies or have any peculiar facility of imagining, would well enough conceive my meaning only by words,' but others required visual assistance. He apologized for the relative poverty of the images, 'being myself absent from the engraver for a good part of the time he was at work, some of the cuts were misplaced, and not graven in the plates.'[26]

Thus, visual representations, few as they necessarily were in Boyle's texts, were mimetic devices. By virtue of the density of *circumstantial* detail that could be conveyed through the engraver's laying of lines, the images imitated reality and gave the viewer a vivid impression of the experimental scene. The sort of naturalistic images that Boyle favoured provided a greater density of circumstantial detail than would have been proffered by more schematic representations. The images served to announce that 'this was really done' and that it was done in the way stipulated; they allayed distrust and facilitated virtual witnessing. Therefore, understanding the role of pictorial representations offers a way of appreciating what Boyle was trying to achieve with his literary technology.[27]

In the introductory pages of the *New Experiments*, Boyle's first published experimental findings, he directly announced his intention to be 'somewhat prolix'. His excuses were three-fold: first delivering things 'circumstantially' would, as we have already seen, facilitate replication; second, the density of circumstantial details was justified by the fact that these were 'new' experiments, with novel conclusions drawn from them: it was therefore necessary that they be 'circumstantially related, to keep the reader from distrusting them'; third, circumstantial reports such as these offered the possibility of virtual witnessing. As Boyle said, 'these narratives [are to be] as standing records in our new pneumatics, and [readers] need not reiterate themselves an experiment *to have as distinct an idea of it*, as may suffice them to ground their reflexions and speculations upon'.[28] If one wrote an experimental report in the

---

25   Hooke to Boyle, 25 August and 8 September 1664, in *RBW*, vol. VI, 487–90, and R. E. W. Maddison, 'The Portraiture of the Honourable Robert Boyle, FRS', *Annals of Science*, vol. 15 (1959), 141–214.

26   Boyle, 'Continuation of New Experiments', op. cit. note 22, 178.

27   Unfortunately, this paper was completed before I was able to read Svetlana Alpers's brilliant *The Art of Describing: Dutch Art in the Seventeenth Century* (London: John Murray; Chicago: The University of Chicago Press, 1983). [. . .]

28   Boyle, 'New Experiments', op. cit. note 1, 1–2 (emphases added). The role of circumstantial detail in Boyle's prose and in that of other early Fellows of the Royal Society is treated in Shapiro, op. cit. note 2, Chapter 7. [. . .]

correct way, the reader could take on trust that these things happened. Further, it would be as if that reader had been present at the proceedings. He would be recruited as a witness and be put in a position where he could validate experimental phenomena as matters of fact.[29] Therefore, attention to the writing of experimental reports was of equal importance to doing the experiments themselves.

In the late 1650s Boyle devoted himself to laying down the rules for the literary technology of the experimental programme. Stipulations about how to write proper scientific prose are dispersed throughout his experimental reports of the 1660s, but he also composed a special tract on the subject of 'experimental essays'. Here Boyle offered extended apologia for his 'prolixity': 'I have,' he understated, 'declined that succinct way of writing'; he had sometimes 'delivered things, to make them more clear, in such a multitude of words, that I now seem even to myself to have in divers places been guilty of verbosity . . .' Not just his 'verbosity' but also Boyle's ornate sentence-structure, with appositive clauses piled on top of each other, was, he said, part of a plan to convey circumstantial details and to give the impression of verisimilitude:

> . . . I have knowingly and purposely transgressed the laws of oratory in one particular, namely, in making sometimes my periods [i.e., complete sentences] or parentheses over-long; for when I could not within the compass of a regular period comprise what I thought requisite to be delivered at once, I chose rather to neglect the precepts of rhetoricians, than the mention of those things, which I thought pertinent to my subject, and useful to you, my reader.[30]

Elaborate sentences, with circumstantial details encompassed within the confines of one grammatical entity, might mimic that immediacy and simultaneity of experience afforded by pictorial representations.

Boyle was endeavouring to constitute himself as a reliable purveyor of experimental testimony and to offer conventions by means of which others could do likewise. The provision of circumstantial details of experimental scenes was a way of assuring readers that real experiments had yielded the findings stipulated. It was also necessary, in Boyle's view, to offer readers circumstantial accounts of *failed* experiments. This performed two functions: first, it allayed anxieties in those neophyte experimentalists whose expectations

29 There is probably a conection between Boyle's justification for circumstantial reporting and Bacon's argument in favour of 'initiative' (as opposed to 'magistral') methods of communication in science: see, for example, D. L. Hodges, 'Anatomy as Science', *Assays*, vol. 1 (1981), 73–89, esp. 83–84; L. Jardine, *Francis Bacon: Discovery and the Art of Discourse* (Cambridge: Cambridge University Press, 1974), 174–78; K. R. Wallace, *Francis Bacon on Communication & Rhetoric* (Chapel Hill, NC: The University of North Carolina Press, 1943), 18–19. [. . .]
30 Boyle, 'Proëmial Essay', op. cit. note 10, 305–06; cf. Boyle, 'New Experiments', op. cit. note 1, 1: R. S. Westfall, 'Unpublished Boyle Papers relating to Scientific Method', *Annals of Science*, vol. 12 (1956), 63–73, 103–17.

of success were not immediately fulfilled; second, it assured the reader that the relator was not wilfully suppressing inconvenient evidence, that he was in fact being faithful to reality. Complex and circumstantial accounts were to be taken as undistorted mirrors of complex experimental performances, in which a wide range of contingencies might influence outcomes.[31] So, for example, it was not legitimate to hide the fact that air-pumps sometimes did not work properly or that they often leaked: '. . . I think it becomes one, that professeth himself a faithful relator of experiments not to conceal' such unfortunate contingencies.[32] It is, however, vital to keep in mind that the contingencies proffered in Boyle's circumstantial accounts represent a selection of possible contingencies. There was not, nor can there be, any such thing as a report which notes all circumstances which might affect an experiment. Circumstantial, or stylized, accounts do not, therefore, exist as pure forms but as publicly acknowledged moves towards or away from the reporting of contingencies.

The Modesty of Experimental Narrative

The ability of the reporter to multiply witnesses depended upon readers' acceptance of him as a provider of reliable testimony. It was the burden of Boyle's literary technology to assure his readers that he was such a man as should be believed. He therefore had to find the means to make visible in the text the accepted tokens of a man of good faith. One technique has just been discussed: the reporting of experimental failures. A man who recounted unsuccessful experiments was such a man whose objectivity was not distorted by his interests. Thus, the literary display of a certain sort of morality was a technique in the making of matters of fact. A man whose narratives could be credited as mirrors of reality was a 'modest man'; his reports should make that modesty visible.

Boyle found a number of ways of displaying modesty. One of the most straightforward was the use of the form of the experimental essay. The essay (that is, the piece-meal reporting of experimental trials) was explicitly con-

31 Boyle, 'Unsuccessfulness of Experiments', op. cit. note 12, 339–40, 353; Recognizing that contingencies might affect experimental outcomes was also a way of tempering inclinations to reject good testimony too readily. If an otherwise reliable authority stipulated an outcome that was not immediately obtained, one was advised to persevere; see ibid., 344–45; Boyle, 'Continuation of New Experiments', op. cit. note 22, 275–76; Boyle, 'Hydrostatical Paradoxes', op. cit. note 11, 743; Westfall, op. cit. note 30, 72–73.
32 Boyle, 'New Experiments', op. cit. note 1, 26. For an example of Boyle reporting an experimental failure, see ibid., 60–70. A critic like Hobbes could capitalize upon Boyle's reported failures, or, more interestingly, deconstruct Boyle's reported successes by identifying further contingencies which affected experimental outcomes; see, for instance, Hobbes, op. cit. note 15, 245–46.

trasted to the natural philosophical system. Those who wrote entire systems were identified as 'confident' individuals, whose ambition extended beyond what was proper or possible. By contrast, those who wrote experimental essays were 'sober and modest men', 'diligent and judicious' philosophers, who did not 'assert more than they can prove.' This practice cast the experimental philosopher into the role of intellectual 'under-builder', or even that of 'a drudge of greater industry than reason'. This was, however, a noble character, for it was one that was freely chosen to further 'the real advancement of true natural philosophy' rather than personal reputation.[33] The public display of this modesty was an exhibition that concern for individual celebrity did not cloud judgement and distort the integrity of one's reports. In this connection it is absolutely crucial to remember who it was that was portraying himself as a mere 'under-builder'. He was the son of the Earl of Cork, and everyone knew that very well. Thus, it was plausible that such modesty could have a noble character, and Boyle's presentation of self as a role model for experimental philosophers was powerful.[34]

Another technique for displaying modesty was Boyle's professedly 'naked way of writing'. He would eschew a 'florid' style; his object was to write 'rather in a philosophical than a rhetorical strain'. This plain, puritanical, unadorned (yet convoluted) style was identified as *functional*. It served to exhibit, once more, the philosopher's dedication to community service rather than to his personal reputation. Moreover, the 'florid' style to be avoided was a hindrance to the clear provision of virtual witness: it was, Boyle said, like painting 'the eye-glasses of a telescope'.[35]

The most important literary device Boyle employed for demonstrating modesty acted to protect the fundamental epistemological category of the experimental programme: the matter of fact. There were to be appropriate moral postures, and appropriate modes of speech, for epistemological items on either side of the crucial boundary that separated matters of fact from the locutions used to account for them: theories, hypotheses, speculations, and the like. Thus, Boyle told his nephew,

in almost every one of the following essays I . . . speak so doubtingly, and use so often, *perhaps, it seems, it is not improbable*, and such other expressions, as argue a diffidence of the truth of the opinions I incline to, and that I should

33 Boyle, 'Proëmial Essay', op. cit. note 10, 300–01, 307; cf. 'Sceptical Chymist', op. cit. note 11, 469–70, 486, 584. Several of the less modest personalities of seventeenth-century English science were individuals who lacked the gentle birth that routinely enhanced the credibility of testimony: e.g., Hobbes, Hooke, Wallis and Newton.
34 The best source for Boyle's social situation and temperament is J. R. Jacob, *Robert Boyle and the English Revolution: A Study in Social and Intellectual Change* (New York: Burt Franklin, 1977), Chapters 1–2.
35 Boyle, 'Proëmial Essay', op. cit. note 10, 318, 304. [. . .]

be so shy of laying down principles, and sometimes of so much as venturing at explications.

Since knowledge of physical causes was only 'probable', this was the correct moral stance and manner of speech, but things were otherwise with matters of fact, and here a confident mode was not only permissible but necessary:

> . . . I dare speak confidently and positively of very few things, except of matters of fact.[36]

It was necessary to speak confidently of matters of fact because, as the foundations of proper philosophy, they required protection. And it was proper to speak confidently of matters of fact, because they were not of one's own making; they were, in the empiricist model, discovered rather than invented. As Boyle told one of his adversaries, experimental facts can 'make their own way' and 'such as were very probable, would meet with patrons and defenders . . .'[37] The separation of modes of speech, and the ability of facts to make their own way, was made visible on the printed page. In *New Experiments* Boyle said he intended to leave 'a conspicuous interval' between his narratives of experimental findings and his occasional 'discourses' upon their interpretation. One might then read the experiments and the 'reflexions' separately.[38] Indeed, the construction of Boyle's experimental essays makes manifest the proper balance between the two categories: *New Experiments* consists of a sequential narrative of 43 pneumatic experiments; *Continuation* of 50; and the second part of *Continuation* of an even larger number of disconnected experimental observations, only sparingly larded with interpretative locutions.

The confidence with which one ought to speak about matters of fact extended to stipulations about the proper use of authorities. Citations of other writers should be employed to use them not as 'judges, but as witnesses', as 'certificates to attest matters of fact.' If this practice ran the risk of identifying the experimental philosopher as an ill-read philistine, it was, however, necessary: '. . . I could be very well content to be thought to have scarce looked upon any other book than that of nature.'[39] The injunction against citing of authorities performed a significant function in the mobilization of assent to matters of fact. It was a way of displaying that one was aware of the workings of the Baconian 'Idols' and was taking measures to mitigate their corrupting effects on knowledge-claims.[40] A disengagement between experimental narrative and

36   Boyle, 'Proëmial Essay', op. cit. note 10, 307 (emphases in original). [. . .]
37   Boyle, 'Hydrostatical Discourse', op. cit. note 11, 596.
38   Boyle, 'New Experiments', op. cit. note 1, 2.
39   Boyle, 'Proëmial Essay', op. cit. note 10, 313, 317.
40   On the 'idols' and fallibilism, see Shapiro, op. cit. note 2, 61–62.

the authority of systematists served to dramatize the author's lack of preconceived expectations and, especially, of theoretical investments in the outcome
of experiments. For example, Boyle several times insisted that he was an inno-
cent of the great theoretical systems of the seventeenth century. In order to
reinforce the primacy of experimental findings, 'I had purposely refrained from
acquainting myself thoroughly with the intire system of either the Atomical,
or the Cartesian, or any other whether new or received philosophy . . .' And,
again, he claimed that he had avoided a systematic acquaintance with the
systems of Gassendi, Descartes, and even of Bacon, 'that I might not be pre-
possessed with any theory or principles . . .'[41]

Boyle's 'naked way of writing', his professions and displays of humility, and
his exhibition of theoretical innocence all complemented each other in the
establishment and the protection of matters of fact. They served to portray the
author as a disinterested observer and his accounts as unclouded and undistorted mirrors of nature. Such an author gave the signs of a man whose testimony was reliable. Hence, his texts could be credited and the number of
witnesses to his experimental narratives could be multiplied indefinitely.

Scientific Discourse and the Community

I have said that the matter of fact was a social as well as an intellectual category. And I have argued that Boyle deployed his literary technology so as to
make virtual witnessing a practical option for the validaton of experimental
performances. I want in this section to examine the ways in which Boyle's
literary technology dramatized the social relations proper to a community of
experimental philosophers. Only by establishing right rules of discourse
between indviduals could matters of fact be generated and defended, and only
by constituting these matters of fact into the agreed foundations of knowledge
could a moral community of experimentalists be created and sustained.
Matters of fact were to be produced in a public space: a particular space
in which experiments were collectively performed and directly witnessed
and an abstract space constituted through virtual witnessing. The problem of
producing this kind of knowledge was, therefore, the problem of maintaining
a certain form of discourse and a certain form of social solidarity. In the following sections I will discuss the ways in which Boyle's literary technology
worked to create and maintain this social solidarity amongst experimental
philosophers.

41   Boyle, 'Some Specimens of an Attempt to Make Chymical Experiments Useful to Illustrate the
Notions of the Corpuscular Philosophy. The Preface', in *RBW*, vol. I, 354–59, at 355 (orig. publ.
1661), Boyle, 'Proëmial Essay', op. cit. note 10, 302. [. . .]

The Linguistic Boundaries of the Experimental Community

In the late 1650s and early 1660s, when Boyle was formulating his experimental and literary practices, the English experimental community was still in its infancy. Even with the founding of the Royal Society, the crystallization of an experimental community centred on Gresham College, and the network of correspondence organized by Henry Oldenburg, the experimental programme was far from securely institutionalized. Criticisms of the experimental way of producing physical knowledge emanated from English philosophers (notably Hobbes) and from Continental writers committed to rationalist methods and to the practice of physics as a demonstrative discipline. Experimentalists were made into figures of fun on the Restoration stage: Thomas Shadwell's *The Virtuoso* dramatized the absurdity of weighing the air, and scored most of its good jokes by parodying the convoluted language of Sir Nicholas Gimcrack (Boyle).[42] The practice of expermental philosophy, despite what numerous historians have assumed, was not overwhelmingly popular in Restoration England.[43] In order for experimental philosophy to be established as a legitimate activity, several things needed to be done. First, it required *recruits*: experimentalists had to be enlisted as neophytes, and converts from other forms of philosophical practice had to be obtained. Second, the social role of the experimental philosopher and the linguistic practices appropriate to an experimental community needed to be defined and publicized.[44] What was the proper nature of discourse in such a community? What were the linguistic signs of competent membership? And what uses of language could be taken as indications that an individual had transgressed the conventions of the community?

The entry fee to the experimental communty was to be the communication of a candidate matter of fact. In *The Sceptical Chymist*, for instance, Boyle extended an olive-branch even to the alchemists. The solid experimental findings produced by some alchemists could be sifted from the dross of their 'obscure' speculations. Since the experiments of the alchemists (and of the Aristotelians) frequently 'do not evince what they are alleged to prove', the former could be accepted into the experimental philosophy by stripping away the theoretical language with which they happened to be glossed. As Carneades (Boyle's mouthpiece) said,

---

42   Shadwell's play was performed in 1676. There is some evidence that Hooke believed *he* was the model for Gimcrack; see R. S. Westfall, 'Hooke, Robert', in *Dictionary of Scientific Biography*, vol. VI, 481–88, at 483. Charles II, the Royal Society's patron, was also said to have found the weighing of the air rather funny.

43   For the extent to which experimental philosophy was, in fact, popular, see Hunter, op. cit. note 35, Chapters 3, 6.

44   This is not intended as an exhaustive catalogue of the measures necessary for institutionalization. Obviously, patronage was required and alliances had to be forged with existing powerful institutions.

... your hermetic philosophers present us, together with divers substantial and noble experiments, theories, which either like peacocks feathers make a great shew, but are neither solid nor useful; or else like apes, if they have some appearance of being rational, are blemished with some absurdity or other, that, when they are attentively considered, make them appear ridiculous.[45]

Thus, those alchemists who wished to be incorporated into a legitimate philosophical community were instructed what linguistic practices could secure their entry. The same principles were laid down with respect to any practitioner: 'let his opinions be never so false, his experiments being true, I am not obliged to believe the former, and am left at liberty to benefit myself by the latter.'[46] By arguing that there was only a contingent, not a necessary, connection between the language of matters of fact and theoretical language, Boyle was defining the linguistic terms upon which existing communities could join the experimental enterprise. They were liberal terms, which might serve to maximize potential membership.[47]

There were other natural philosophers Boyle despaired to recruit. Hobbes, notably, was the kind of philosopher who, on no account, ought to be admitted, for he denied the value of systematic and elaborate experimentation, the foundational status of the matter of fact, and the distinction between causal and descriptive language. Of Hobbes's *Dialogus physicus*, Boyle asked 'What new experiment or matter of fact Mr *Hobbes* has therein added to enrich the history of nature . . .?' In his criticisms of Boyle's experiments Hobbes 'does not, that I remember, deny the truth of any of the matters of fact I have delivered.' According to Boyle, both Hobbes and another critic, the Jesuit Franciscus Linus, had not 'seen cause to deny any thing that I deliver as experiment.'[48] One could not be regarded as a competent member of the experimental community if one failed to communicate experimental matters of fact, or if one did so in a manner that failed to recognize the linguistic boundaries between factual and causal locutions.

Linguistic Boundaries within the Experimental Community

Just as linguistic categories were used to manage entry to the experimental community, distinctions between the language of facts and that of theories

45 Boyle, 'Sceptical Chymist', op. cit. note 11, esp. 468, 513, 550, 584.
46 Boyle, 'Proëmial Essay', op. cit. note 10, 303.
47 Boyle's way of dealing with the hermetics drew on the views of the Hartlib group of the 1640s and 1650s. By contrast, there were those who rejected the findings of late alchemy (e.g., Hobbes) and those who rejected the process of assimilation (e.g., Newton).
48 Boyle, 'Examen of Hobbes', op. cit. note 7, 233, 197; Boyle, 'Defence against Linus', op. cit. note 11, 122.

were deployed to regulate discourse within it. In broad terms, Boyle insisted upon a separation between 'physiological' and 'metaphysical' languages: experimental discourse was to be confined to the former. One of the central categories of Boyle's 'new pneumatics' also happened to be a major preoccupation of the old physics – namely, vacuism versus plenism, and the judgement whether a vacuum was possible in nature. How was it proper to speak of the contents of the receiver of an evacuated air-pump? And how did this speech relate to traditional usages of the term 'vacuum'?

A practical problem was posed by the fact that the lexicon of the new philosophy was largely compiled out of the usages of old discursive practices. Old words had to be given new meanings. Thus, it was proper to apply the term 'vacuum' to the contents of the exhausted receiver, but it was improper to take this to mean that the space was absolutely devoid of all matter. Such an absolutely void space was the 'vacuum' of metaphysical discourse. What Boyle meant by the air-pump's 'vacuum' was 'not a space, wherein there is no body at all, but such as is either altogether, or almost totally devoid of air.'[49] If contemporary plenists maintained that this vacuum might be filled by a subtle form of matter, or 'aether', Boyle could reply with a series of experiments which showed that such an aether could not be made 'sensible', that is, it had no physical manifestations. And speech of entities that were not amenable to sensible experimentation was not permissible within experimental philosophy.[50]

The separation of 'physiological' from 'metaphysical' language was most crucial to Boyle's strategy for dealing with causal inquiry in physical science. In keeping with his probabilist conception of knowledge, Boyle wished to bracket off speech about matters of fact, about which one might be certain, from speech of their physical causes, which were at best probable. In terms of Boyle's air-pump programme, the most important instance of this bracketing concerned the notion which was the main product of these experiments: the 'spring of the air'. Boyle said that his 'business' was 'not to assign the adequate cause of the spring of the air, but only to manifest, that the air hath a spring, and to relate some of its effects.' The cause of the air's elasticity *might* be accounted for variously: by Cartesian vortices, or by the real physical existence in the corpuscles of the air of 'slender springs' or of a fleecy structure.[51] The job of the experimental philosopher was to speak of

49   Boyle, 'New Experiments', op. cit. note 1, 10.

50   Boyle, 'Continuation of New Experiments', op. cit. note 22, 250–58. Note that in other contexts Boyle encouraged speech of immaterial entities such as spirits; what he said was that such items ought to be purged from the routine discourse of experimental philosophy; see, for example, 'Hydrostatical Discourse', op. cit. note 11, 608.

51   Boyle, 'New Experiments', op. cit. note 1, 11–12; cf. Boyle, 'The General History of the Air . . .', in *RBW*, vol. V, 609–743, at 614–15 (orig. publ. 1692).

experimentally-produced matters of fact, not to conjecture further than that.[52]

Boyle had considerable problems in difffusing this new mode of speech. Plenist critics persisted in understanding Boyle to be using 'vacuum' in its metaphysical sense, and Boyle was obliged persistently to reiterate its proper usage.[53] Other writers either refused to conceive of a natural philosophy that bracketed off causal speech, or reckoned that Boyle must be committed to some (illegitimate and unacknowledged) causal account of the spring of the air.[54] So far as the 'spring of the air' was concerned, Boyle's stipulation that it had been made experimentally 'manifest' and his disinclination to speak of its cause had an interesting effect. By putting the spring on the other side of the boundary from causal locutions, Boyle constituted the spring, for all practical purposes, into a matter of fact. When it came to labelling the epistemological status of the spring, Boyle variously referred to it as an 'hypothesis' or even as a 'doctrine'. However, by making the spring into something that was made manifest through experiment, and by protecting it from the uncertainties that afflicted epistemological items like causal notions, Boyle treated this 'hypothesis' in the same way that he treated other matters of fact.[55]

The vital difference between matters of fact and all other epistemological categories was the degree of assent one might expect to them. To an authenticated matter of fact all men will assent. In Boyle's system that was taken for granted because it was through the technologies that multiplied witness that matters of fact were constituted. General assent was what made matters of fact, and general assent was therefore mobilized around matters of fact. With 'hypotheses', 'theories', 'conjectures', and the like, the situation was quite different. These categories threatened that assent which could be crystallized in the institution of the matter of fact. Thus, the linguistic conventions of Boyle's experimental programme separated speech appropriate to the two categories as a way of drawing the boundaries between that about which one was to expect certainty and assent and that about which one could expect uncertainty and divisiveness. The idea was not to eliminate dissent or to oblige men to agree to all items in natural philosophy (as it was for Hobbes); rather, it was

---

52   These problems were structurally similar to those afflicting Newton later in the century. Newton said that he wished to speak of gravitation as a mathematical regularity, without venturing an account of its physical cause. Newton's allies and enemies alike found it difficult to accept such mathematical statements as the end-product of physical inquiry; see A, Koyré, *Newtonian Studies* (Chicago: The University of Chicago Press, 1968), 115–63, 273–82.

53   Boyle, 'Defence against Linus', op. cit. note 11, 135, 137; Boyle, 'Examen of Hobbes', op. cit. note 7, 191, 207; Boyle, 'Animadversions on Hobbes', op. cit. note 7, 112.

54   Hobbes, op. cit. note 15, 271, 273, 278; for Boyle's reply, see 'Examen of Hobbes', op. cit. note 7, 193–94.

55   Cf. Boas [Hall], 'Boyle', op. cit. note 41; Boas, 'Establishment', op. cit. note 41, 475–77.

to manage dissent and to keep it within safe bounds. An authenticated matter of fact was treated as a mirror of nature; a theory, by contrast, was clearly man-made and could, therefore, be contested. Boyle's linguistic boundaries acted to segregate what could be disputed from what could not. The management of dispute in experimental philosophy was crucial to protecting the foundations of knowledge.

Manners in Dispute

Since natural philosophers were not to be compelled to give assent to all items of knowledge, dispute and controversy was to be expected. How should this be dealt with? The problem of conducting dispute was a matter of intense practical concern in early Restoration science. During the Civil War and Interregnum the divisiveness of 'enthusiasts', sectarians and hermeticists threatened to bring about radical individualism in philosophy. Nor did the various sects of Peripatetic natural philosophers display a public image of a stable and united intellectual community. Unless the new experimental community could exhibit a broadly-based consensus and harmony within its own ranks, it was unreasonable to expect it to secure the legitimacy within Restoration culture that its leaders desired. Moreover, that very consensus was vital to the establishment of matters of fact as the foundational category of the new practice.

By the early 1660s Boyle was in a position to give concrete exemplars of how disputes ought to be conducted; three critics published their responses to his *New Experiments*, and he replied to each one: Linus, Hobbes and Henry More. But even before he had been engaged in dispute, Boyle laid down a set of rules for how controversies were to be handled by the experimental philosopher. For example, in *A Proëmial Essay* (composed 1657), Boyle insisted that disputes should be about findings and not about persons. It was proper to take a hard view of reports which were inaccurate but most improper to attack the character of those who rendered them: 'for I love to speak of persons with civility, though of things with freedom'. The *ad hominem* style must at all costs be avoided, for the risk was that of making foes out of mere dissenters. This was the key point: potential contributors of matters of fact, however wrong they may be, must be treated as possible converts to the experimental philosophy. If, however, they were bitterly treated, they would be lost to the cause and to the community whose size and consensus validated matters of fact:

> And as for the (very much too common) practice of many, who write, as if they thought railing at a man's person, or wrangling about his words, necessary to the confutation of his opinions; besides that I think such a quarrelsome and

injurious way of writing does very much misbecome both a philosopher and a Christian, methinks it is as unwise, as it is provoking. For if I civilly endeavour to reason a man out of his opinions, I make myself but one work to do, namely, to convince his understanding; but, if in a bitter or exasperating way I oppose his errors, I increase the difficulties I would surmount, and have as well his affections against me as his judgment: and it is very uneasy to make a proselyte of him, that is not only a dissenter from us, but an enemy to us.[56]

Furthermore, it was impolitic to acknowledge the existence of 'sects' in natural philosophy. One way by which one could hope to overcome sectarianism was to decline public recognition that it existed: 'it is none of my design,' Boyle said, 'to engage myself with, or against, any one sect of Naturalists . . .' The experiments will decide the case. The views of these 'sects' should be noted only insofar as they are founded upon experiment. Therefore, it was right and politic to be harsh in one's writings against those who do not contribute experimental findings, for they have nothing to offer to the constitution of matters of fact. Finally, the experimental philosopher must show that there was point and purpose to legitimately conducted dispute. He should be prepared publicly to renounce positions that were shown to be erroneous. Flexibility followed from fallibilism. As Boyle wrote, 'till a man is sure he is infallible, it is not fit for him to be unalterable.'[57]

The conventions for managing dispute were dramatized in the structure of *The Sceptical Chymist*. These fictional conversations (between an Aristotelian, two varieties of hermeticists, and 'Carneades' as mouth-piece for Boyle) took the form, not of a Socratic dialogue, but of a *conference*.[58] They were a little piece of theatre that exhibited how persuasion, dissensus and, ultimately, conversion to truth ought to be conducted. Several points about Boyle's theatre of persuasion can be briefly made: first, the 'symposiasts' are imaginary, not real. This means that opinions can be confuted without exacerbating relations between real philosophers. Even Carneades, although he is manifestly 'Boyle's man', is not Boyle himself: Carneades is made actually to quote 'our friend Mr *Boyle*' as a device for distancing opinions from individuals. The author is insulated from the text and from the opinions he may actually espouse. Second, truth is not inculcated from Carneades to his interlocutors; rather it is dramatized as emerging through the conversation.[59] Everyone is seen to have a say

56   Boyle, 'Proëmial Essay', op. cit. note 10, 312.
57   Ibid., 311.
58   See R. P. Multhauf, 'Some Nonexistent Chemists of the Seventeenth Century: Remarks on the Use of the Dialogue in Scientific Writing', in A. G. Debus and Multhauf, *Alchemy and Chemistry in the Seventeenth Century* (Los Angeles, Calif.: William Andrews Clark Memorial Library, 1966), 31–50.
59   Boyle, 'Sceptical Chymist', op. cit. note 11, 486. In the preface Boyle says that he will not 'declare my own opinion'; he wishes to be 'a silent auditor of their discourses' (460, 466–67).

in the consensus which is the dénouement.[60] Third, the conversation is, without exception, civil: as Boyle said, 'I am not sorry to have this opportunity of giving an example, how to manage even disputes with civility . . .'[61] No symposiast abuses another; no ill temper is displayed; no one leaves the conversation in pique or frustration.[62] Fourth, and most importantly, the currency of intellectual discourse, and the means by which agreement is reached, is the experimental matter of fact. Here, as I have indicated, matters of fact are not treated as the exclusive property of any one philosophical sect. Insofar as the alchemists have produced experimental findings, they have minted the real coins of experimental exchange. Their experiments are welcome, while their 'obscure' speculations are not. Insofar as the Aristotelians produce few experiments, and insofar as they refuse to dismantle the 'arch'-like 'mutual coherence' of their philosophical system into facts and theories, they can make little contribution to the experimental conference.[63] In these ways, the structure and the linguistic conventions of this imaginary conversation make vivid the rules for real conversations proper to experimental philosophy. [. . .]

Scientific Knowledge and Exposition: Conclusions

I have shown that three technologies were involved in the production and validation of Boyle's experimental matters of fact: the material, the literary and the social. Although I have concentrated here upon the literary technology, I have also suggested that the three technologies are not distinct: the working of each depends upon and incorporates the others. I want now briefly to develop that point by showing how each technology contributes to a common strategy for constituting matters of fact.

What makes a fact different from an artefact is that the former is not perceived to be man-made. What men make, men may unmake, but a matter of fact is taken to be the very mirror of nature. To identify the role of human agency in the making of an item of knowledge is to identify the possibility of its being otherwise. To shift agency on to natural reality is to stipulate the grounds for universal assent. Each of the three technologies works to achieve the appearance of matters of fact as *given* items: each functions as an objectifying resource.

60   The consensus that emerges is very like the position from which Carneades starts, but the plot of *The Sceptical Chymist* involves disguising that fact. [. . .]
61   Boyle, 'Sceptical Chymist', op. cit. note 11, 462.
62   Actually, the great bulk of the talk is between Carneades and Eleutherius. The other two participants inexplicably absent themselves from most of the proceedings. This is possibly an accident due to Boyle's self-confessed sloppiness with his manuscripts; he was continually apologizing for losing pages of his drafts.
63   Boyle, 'Sceptical Chymist', op. cit. note 11, 469.

Take, for example, the role of the air-pump in the production of matters of fact. As I have noted, pneumatic facts were machine-made. The product of the pump was not, as it is for the modern scientific machines studied by Latour, an 'inscription': it was a visual experience that had to be transformed into an inscription by a witness.[64] However, the air-pump of the 1660s has this in common with the gamma counter of the present-day neuroendocrinological laboratory: it stands between the perceptual competences of a human being and natural reality itself. A 'bad' observation taken from a machine need not be ascribed to cognitive or moral faults in the human being, nor is a 'good' observation his personal product. It is the machine that has generated the finding. A striking instance of this usage arose in the 1660s when Christiaan Huygens offered a matter of fact produced by his pump which appeared to conflict with one of Boyle s central explanatory resources. Boyle did not impugn Huygens's integrity or his perceptual and cognitive competences. Instead, he suggested that the fault lay with the machine: '[I] question not his Ratiocination, but only the staunchness of his pump.'[65] The machine constitutes a resource that may be used to factor out human agency in the intellectual product: 'it is not I who says this; it is the machine that speaks,' or 'it is not your fault; it is the machine's.'

Boyle's social technology constituted an objectifying resource by making the production of knowledge visible as a collective enterprise: 'it is not I who says this; it is all of us.' As Sprat insisted, collective performance and collective witness served to correct the natural working of the 'idols': the faultiness, the idiosyncracy or the bias of any individual's judgement and observational ability. The Royal Society advertised itself as a 'union of eyes, and hands'; the space in which it produced its experimental knowledge was stipulated to be a *public space*. It was public in a very precisely defined and very rigorously policed sense: not everyone could come in; not everyone's testimony was of equal worth; not everyone was equally able to influence the official voice of the institution. Nevertheless, what Boyle was proposing, and what the Royal Society was endorsing, was a crucially important *move towards* the public constitution and validation of knowledge. The contrast was, on the one hand, with the private work of the alchemists, and, on the other, with the individual dictates of the systematical philosophers.

In the official formulation of the Royal Society, the production of experimental knowledge commenced with individuals' acts of seeing and believing,

64   B. Latour and S. Woolgar, *Laboratory Life: The Social Construction of Scientific Facts* (Beverly Hills, Calif.: Sage, 1979), Chapter 2; and, for a fine study of the role of instruments in scientific observation reports, see T. J. Pinch, 'Towards an Analysis of Scientific Observation: The Externality and Evidential Significance of Observational Reports in Physics', *Social Studies of Science*, vol. 15 (1985), in press.
65   Boyle to R. Moray, July 1662, in Huygens, op. cit. note 9, vol. IV, 217–20; cf. Boyle, 'Defence against Linus', op. cit. note 11, 152–53.

and was completed when all individuals voluntarily agreed with one another about what had been seen and ought to be believed. This freedom to speak had to be protected by a special sort of discipline. Radical individualism – each individual setting himself up as the ultimate judge of knowledge – would destroy the conventional basis of knowledge, while the disciplined collective social structure of the experimental language game would create and sustain that factual basis. Thus, the experimentalists were on guard against 'dogmatists' and 'tyrants' in philosophy, just as they abominated 'secretists' who produced their knowledge-claims in a private space. No one man was to have the right to lay down what was to count as knowledge. Legitimate knowledge was objective insofar as it was produced by the collective, and agreed to voluntarily by those who comprised the collective. The objectification of knowledge proceeded through displays of the communal basis of generation and evaluation. Human coercion was to have no visible place in the experimental way of life.[66]

It was the function of the literary technology to create that communal way of life, to bound it, and to provide the forms and conventions of social relations within it. The literary technology of virtual witnessing supplemented the public space of the laboratory by extending a valid witnessing experience to all readers of the text. The boundaries stipulated by Boyle's linguistic practices acted to keep that community from fragmenting and served to protect items of knowledge to which one could expect universal assent from items which produced divisiveness. Similarly, Boyle's stipulations concerning proper manners in dispute worked to guarantee that social solidarity which generated assent to matters of fact and to rule out of order those imputations which would undermine the moral integrity of the experimental way of life.

I have attempted to display these linguistic practices in the making, and, within restrictions of space, I have alluded to sources of seventeenth-century opposition to these practices. It is important to understand two things about these ways of expounding scientific knowledge and securing assent: that they are historical constructions and that there have been alternative practices. It is particularly important to understand this because of the problems of givenness and self-evidence that attend the institutionalization and conventionalization of these practices. Just as the three technologies operate to create the illusion that matters of fact are not man-made, so the institutionalized and conventional status of the scientific discourse that Boyle helped to produce makes the illusion that scientists' speech about natural reality is simply a reflection of that reality. In this instance, and in others like it, the historian has two major tasks: to display the man-made nature of scientific knowledge, and to account for the illusion that this knowledge is *not* man-made. It is one of the

66   Sprat, op. cit. note 14, 85 (for 'eyes and hands'), 98–99 (for the individual and the collective), 28–32 (for 'tyrants' in philosophy). [. . .]

recommendations of the sociology of knowledge perspective that analysts often attempt to accomplish these two tasks in the same exercise.[67]

In the late twentieth century scientific papers are rarely, if ever, written with the depth of circumstantial detail which Boyle's reports contained. Why might this be? The answer to this question leads us to the study of linguistic aspects of scientific institutionalization and differentiation. In discussing the characteristics of a *Denkkollektiv*, Ludwik Fleck noted that such a group cultivates 'a certain exclusiveness both formally and in content':

> A thought commune becomes isolated formally, but also absolutely bonded together, through statutory and customary arrangements, sometimes a separate language, or at least special terminology . . . The optimum system of a science, the ultimate organization of its principles, is completely incomprehensible to the novice [or, Fleck might have added, to any non-member].[68]

Fleck was suggesting that the linguistic conventions of a body of practitioners constitute an answer to the question 'Who may speak?' The language of an institutionalized and specialized scientific group is removed from ordinary speech, and from the speech of scientists belonging to another community, both as a sign and as a vehicle of the group's special and bounded status. Not everyone may speak; the ability to speak entails the mastering of special linguistic competences; and the use of ordinary speech is taken as a sign of non-membership and non-competence. Such a group gives linguistic indications that the generation and validation of its knowledge does not require the mobilizing of belief, trust and assent outwith its own social boundaries. (Yet, when external support or subvention is required, special *occasional* modes of speech may be resorted to, including the various languages of 'popularization'.)

By contrast, Boyle's circumstantial reporting was a means of involving a wider community and soliciting its participation in the making of factual experimental knowledge. His circumstantial language was a way of bringing readers into the experimental scene, indeed of making the reader an actor in that scene. The reader was to be shown not just the products of experiments but their mode of construction and the contingencies affecting their performance, *as if he were present*. Boyle aimed to accomplish this, not by inventing

---

67   See especially the work of Collins whose metaphor of completed and consensual scientific knowledge as 'the ship in the bottle' nicely crystallizes this point: for example, H. M. Collins, 'The Seven Sexes: A Study in the Sociology of a Phenomenon, or the Replication of Experiments in Physics', *Sociology*, vol. 9 (1975), 205–24; Collins, 'Son of Seven Sexes: The Social Destruction of a Physical Phenomenon', *Social Studies of Science*, vol. 11 (1981), 33–62. [. . .]
68   L. Fleck, *Genesis and Development of a Scientific Fact*, trans. F. Bradley and T. J. Trenn, eds Trenn and R. K. Merton (Chicago: The University of Chicago Press, 1979), 103, 105 (orig. publ. in German, 1935).

a totally novel language (although it was novel to the natural philosophical community of the time), but, it could be argued, by incorporating aspects of ordinary speech and lay techniques of validating knowledge-claims. The language of early Restoration experimental science was, in this sense, a public language. And the use of this public language was, in Boyle's work, essential to the creation of both the knowledge and the social solidarity of the experimental community. Trust and assent had to be won from a public that might crucially deny trust and assent.

4

The Mechanical Philosophy and Its Appeal

A Mechanical Microcosm: Bodily Passions, Good Manners, and Cartesian Mechanism

Peter Dear

Originally appeared as "A Mechanical Microcosm: Bodily Passions, Good Manners, and Cartesian Mechanism," in *Science Incarnate: Historical Embodiments of Natural Knowledge*, edited by Christopher Lawrence and Steven Shapin (Chicago and London: University of Chicago Press, 1998):51–82.

The text and notes in this chapter have been abridged. For complete footnotes the original publication should be referred to.

Editor's Introduction

Why do people believe or reject scientific theories? Traditionally, the history of science answered this question in purely cognitive terms. Good theories accord well with observations, they predict phenomena accurately, they have great explanatory economy and power. But, as historians have become increasingly aware of the role of social and cultural influences in science, they have expanded their explanations of what makes a theory attractive to include factors which are not purely rational or cognitive. Thomas Kuhn noted Copernicus's appeal to aesthetics in defending his heliocentric theory.[1] Kepler was convinced of the truth of Copernicanism because it accorded well with his Neoplatonism. People accept certain theories because they help them to make sense of the world and accord well with how they think the world should be. Also particular theories can be appealing because they enable people to act in the world or condone or justify certain actions. So the question of why particular theories are accepted is certainly a complex one.

---

1 Thomas S. Kuhn, *The Copernican Revolution: Planetary Astronomy in the Development of Western Thought* (Harvard: Harvard University Press, 1957).

Hooykaas states that the spread of mechanical philosophies of nature was one of the most important elements of the Scientific Revolution. Here Peter Dear attempts to account for the great popularity of René Descartes's version of the mechanical philosophy in the seventeenth century, not just among philosophers but throughout large sections of society in France and in the Netherlands. What was the appeal of Descartes's mechanical world view? To understand it, Dear argues, we need to see how natural philosophy met the demands of early modern society. In doing so, he gives us a picture of the natural philosopher which is very different from our modern scientist wearing a white lab coat.

Drawing on the work of Nobert Elias on the civilizing process, Dear suggests that in absolutist society that was increasingly regimented, ordered, and regulated, people who wished to succeed needed to be able to control their passions. For them Descartes's theory of psychology and physiology was particularly useful. According to Descartes's dualistic theory, the human mind or soul was radically distinct from the body. The body was mere matter in motion. In fact, animals, which had no soul, were simply robots or mechanical automata. Humans did have an immaterial soul, but the body could act on it through the pineal gland to create the passions. On the other hand, the soul could rule over the body and subdue and regulate the passions, something of great utility to people who sought to adhere to the civilized norms of behavior valued at the time, or to people such as the exiled Princess Elizabeth of Bohemia overcome with melancholy at her hard lot.

A Mechanical Microcosm: Bodily Passions, Good Manners, and Cartesian Mechanism

Peter Dear

> *Of fencing experts he remarked that they were masters of a science or art which when they needed it they did not know how to employ, adding that there was something presumptuous in their seeking to reduce to infallible mathematical formulas the angry thoughts and impulses of their adversaries.*
>
> Cervantes, *El licenciado vidriera*.[1]

Introduction

The 1669 edition of Sébastien Le Clerc's *Pratique de la geometrie* contains numerous plates designed to augment its instruction to young gentlemen in the mathematical arts. It presents geometry as a practical subject that will assist in such areas as military science and fortification, traditional branches of mixed mathematics in the seventeenth century.[2] It also invokes swordplay as an illustration of the rootedness of geometrical curves in practical operations. The corresponding plate [. . .] displays at the foot of the page four men engaged in combat; the upper half is occupied by geometrical diagrams that display the liens and arcs traced out by the properly wielded blade.[3]

Half a century before, a young French gentleman, already admitted into the culture to which Le Clerc's treatise was to cater, thought to contribute to its refinement by himself writing a manual on fencing. Fencing, like dancing, was one of the basic social accomplishments expected of a nobleman, and as such it demanded a disciplined treatment. This particular manual has been lost, but there can be little doubt that its author, René Descartes, appreciated the value of "infallible mathematical formulas" in civilizing an art so much associated

---

1 Cervantes, *El licenciado vidriera*, 789. I am informed by Dale Pratt of Brigham Young University that such satire of contemporary fencing manuals can also be found in several works by Cervantes's contemporary Francisco de Quevedo, including *La vida del Buscòn*, which refers specifically to Luis Pacheco de Narváez, *Grandezas de las espadas* (1600).

2 Useful accounts of mathematical education and its topical scope in French Jesuit colleges – which trained gentlemen such as Descartes – appear in Dainville, "L'enseignement des mathématiques," and idem, *La géographie des humanistes*, chap. 1. Motley, *Becoming a French Aristocrat*, chap. 3, examines the academy in this period and its topical focus on practical mathematical arts and gentlemanly accomplishments.

3 The plate may be found reproduced in Harth, *Ideology and Culture*, 254; Le Clerc's treatise is discussed on 251–57. I have used the 1682 edition, Le Clerc, *Pratique de la geometrie*.

with "angry thoughts and impulses."[4] Much the same features were found also in the equally gentlemanly, or noble, arts of horsemanship, as contemporary treatises again show. Dressage involved careful, disciplined management of the horse by its rider, but in this case the distance between formalized maneuver and the "angry impulses" of battle was more evident: dressage was dancing on horseback, much as, perhaps, fencing was dancing with swords. Each mimicked violence while in practice eviscerating its affective core, a hypocrisy that Cervantes effectively skewered. Gentlemanly fencers, dancers, and horsemen were automata, going through the motions.[5]

In 1649 Descartes published his *Treatise on the Passions of the Soul*. In it he discusses the influence on the mind of various physiological disturbances, together with observations on the ways in which desirable and undesirable effects may be controlled. No mathematical formulas appear in the course of the exposition, but his mechanistic ontology, the physical instantiation in Descartes's philosophy of mathematical reasoning, underpins the entire discussion. The current chapter is an attempt at making sense of a particular relationship – the Cartesian – between norms of bodily behavior and criteria of intelligibility in seventeenth-century natural philosophy. It does so by investigating the mechanization of natural philosophy and the way in which the resultant structure of intelligible nature constrained (or expressed) the structure of human behavior so intimately connected to it. For Descartes, human behavior was part of human physiology, and that physiology, like the rest of the physical universe, was mechanistic.

Descartes and the Rules of Behavior

René Descartes did not pursue an entirely conventional career for someone of his sort. He was born in 1596, the oldest son of a lawyer, a *conseiller* at the *parlement* of Rennes, who was aiming at the *noblesse de robe*, a very minor bour-

---

4   For other examples of geometrical illustrations of fencing behavior, see the plates, taken from Girard Thibault, *Académie de l'espée* (1628), in Vigarello, "The Upward Training of the Body," 160–65; these pictures integrate the kinds of moves lampooned by Cervantes with familiar Renaissance geometrical overlays on the bodies of the fencers themselves, representing their ideal bodily proportions (for the early seventeenth century, one might point to many examples of the latter in Robert Fludd's works – see Godwin, *Robert Fludd*). See also Motley, *Becoming a French Aristocrat*, 139, on fencing and "the display of the body as a social symbol." On Descartes's fencing treatise see Descartes, *Oeuvres*, 10:533–38. Charles Adam refers to Descartes's probable fencing instruction at La Flèche (as one of the accomplishments of a gentleman) in his "Vie de Descartes," ibid. 12:28.

5   On horsemanship and its pale shadowing of the arts of a former warrior class in the new centralizing French state, see Apostolidès, *Le roi-machine*, 45. See also Motley, *Becoming a French Aristocrat*, 150, on fencing and its increasing stress on "mastering posture and movement of the body."

geois kind of nobility. Around 1606 he entered the recently founded Jesuit college of La Flèche, already one of the most celebrated schools in France, to study classics, rhetoric, the philosophy of Aristotle, and mathematics. He was there until about 1615, receiving an education that was second to none in Europe at that time, and also receiving reinforcement of his own sense of self – the self of a young French gentleman.[6]

Descartes's treatment at La Flèche took full account of his gentle status. Ordinary scholarship pupils from relatively poor families were consigned to shared dormitories and a strict regime that included effectively constant surveillance; whereas those fewer pupils of a higher status were typically permitted private bedrooms and even, in some cases, a private valet. Neither were they required to attend all the classes prescribed for the others. Descartes had no valet, but he made full use of his private sleeping arrangements. On his arrival, he was put under the wing of one Father Charlet, whose concern for his charge extended to allowing him to remain in bed until quite late in the morning (throughout his life Descartes preferred to rise at around ten o'clock). This was, according to Descartes's seventeenth-century biographer Baillet, in part out of deference to René's (admittedly) delicate health, but also a concession to his intellectual propensities, which supposedly lent themselves naturally to morning meditation.[7] In reality, of course, such indulgence flowed more from Descartes's social position than from a Californian concern for personal growth.

According to Baillet's biography, Descartes's father, Joachim, wanted young René, as well as his less intractable brothers, to follow in his footsteps by entering the legal profession. René actually received a law degree and license at Poitiers in 1616 (although it is unclear whether much more was involved than paying the fee), but he did nothing further in that line. René had been destined, after his departure from La Flèche, to go and serve for a spell in the king's army, but Joachim decided that his son was too young and too weak of constitution for such a life. Baillet says that René was therefore sent to sample life in Paris instead.[8]

Evidently, he soon perked up in the capital: Descartes left France in 1618 for Breda in the United Provinces, and the army of Prince Maurice of Nassau. Military adventure was a known option for young French gentlemen, and Descartes's choice of it, despite his explanations in the *Discourse on Method* of

6   Gaukroger, *Descartes*, chap. 2, gives a valuable synthetic biographical reconstruction of Descartes's early life based on what little reliable evidence exists. Rodis-Lewis, "Descartes' Life," is a useful short overview of Descartes's career.

7   Baillet, *Vie*, 1:18, 28. On the background to Baillet's book, see Sebba, "Adrien Baillet," esp. 48–57. Snyders, *La pédagogie en France*, discusses the surveillance aspect of Jesuit colleges in this period. For a brief discussion of "meditation" in nondevotional Jesuit pedagogical strategy see Dear, "Mersenne's Suggestion."

8   Baillet, *Vie*, 1:35. See Gaukroger, *Descartes*, 64–65, on the legal excursion.

1637 having to do with the philosophical and moral benefits of learning from experience, looks as much the product of idleness as of vocation.[9] Descartes seems never to have wanted for money. He had his share of the family estates in Poitou, and that seems to have been quite enough to keep him comfortable.[10] He apparently left the rigors of their administration to his brother and other surrogates; throughout his period of residency in the Netherlands (1628–49) he made only occasional visits to France to deal with business matters.

Descartes's father purportedly remarked, following the publication of the *Discourse*: "Only one of my children has given me displeasure. How can I have given birth to a son silly enough to have himself bound in calf!"[11] Joachim Descartes probably never became reconciled to René's choice of letters as his métier. It was very much an occupation associated with clergymen and scoundrels rather than with gentlemen of families aspiring to the *noblesse de robe*.[12] Descartes was making up his own persona, rather than adopting one ready-made.[13] Descartes's apparent contentment with a relatively modest income led Baillet to remark that "it wasn't at all like a needy and grasping gentleman, but like a rich and content philosopher that M. Descartes regarded the goods of the earth."[14] The model of a philosopher here is evidently an antique one, a Stoic or Epicurean, for example, rather than one based on contemporary schoolmen.

Until his permanent move from France to Holland in 1628, Descartes traveled intermittently, visiting Germany and Italy as well as spending extended periods of time in Paris. He spent two years in Paris before going to live in the Netherlands; his claimed reasons for leaving, at a time when he was starting to devote himself seriously to work in philosophy and mathematical sciences, are important in understanding his ongoing process of self-creation. Descartes said that he had gone to the Low Countries in order to be able to control his time; in Paris he had too many visitors and too many social obligations.[15]

Steven Shapin has written of the unusual steps taken by Robert Boyle in the 1660s and '70s to establish a refuge of privacy for himself by means of which, at particular times, he would make himself unavailable to visitors so

9   Descartes, *Oeuvres* 6:9.
10   Baillet, *Vie*, 2:459, claims that Descartes, through most of his life, lived on between about six and seven thousand livres of annual rental income.
11   Gaukroger, *Descartes*, 20–23, on Descartes's relationship with his father; quote translated on 23.
12   See, in addition to the previous note, Rodis-Lewis, *L'oeuvre de Descartes*, 1:24–25. The rather déclassé contemporary aura of publishing is noted in Thoren, *The Lord of Uraniborg*, 63. The unorthodoxy of Descartes's career path is noted in Sutton, *Science for a Polite Society*, 59.
13   Compare Goffman, *The Presentation of Self in Everyday Life*, with Greenblatt, *Renaissance Self-Fashioning*.
14   Baillet, *Vie*, 2:459.
15   Descartes, *Oeuvres*, 6:31.

as to be able to pursue his natural-philosophical work undisturbed. Boyle's high social status and pan-European reputation as an experimental philosopher, one who might be expected to have many curiosities to show to his callers, rendered him tempting game for those who had themselves the social standing appropriate for imposing on him unexpectedly. Ordinarily it would have been a serious breach of conduct for a gentleman to refuse to see a qualified caller merely because he was otherwise engaged, but Boyle was able to transcend that norm to a limited extent both because of his particularly high noble status and because he artfully drew on legitimating models of reclusive behavior associated with a religious calling.[16] Descartes seems to have encountered the same problems without having available a comparable solution; instead, he fled.[17]

In effect, Descartes presented the move as his own solution to the same problem as that faced later by Boyle. Vocationally, Descartes saw himself as a philosopher, which, outside a school setting, was an odd thing to be. But the philosopher's role did immediately invoke certain generally understood models of behavior – Baillet explained Descartes's initial decision to leave France by saying that Descartes wanted to be able to enjoy solitude, like all great philosophers, who abandon the courts of princes to enjoy study and meditation away from their own country. Three decades or so earlier, the astronomer Tycho Brahe had provided a similar self-presentation in an account of his own life, describing how he had been obliged to go to a foreign land so as to escape the social shackles that prevent peaceful devotion to one's philosophical calling.[18]

But Descartes's life in the Low Countries, seen in relation to this commonplace picture, presents incongruities. He was not in fact the isolated recluse, the solitary thinker, that intellectual mythography has often suggested and that his successfully promulgated self-image portrayed. He was no more freed of social obligations and the entertaining of visitors in his new home than he would have been if located just a few miles outside Paris.[19] For the most part he lived, during his twenty years' Dutch residence, in major towns (Amsterdam, Utrecht, Leiden), and in those places he hobnobbed with socially elevated people who were the local equivalents of those he had left behind; among others, Constantijn Huygens, the father of Christiaan Huygens and a

16   Shapin, "The House of Experiment"; idem, " 'The Mind Is Its Own Place.' "
17   It may be valuable to consider by contrast the case of Descartes's popular philosophical friend Mersenne (although he was of humble social origin): he would have been able to avoid callers when he wished because he had religious duties giving him a legitimate claim to solitude.
18   Hannaway, "Laboratory Design and the Aim of Science," 590–91; Thoren, *The Lord of Uraniborg*, 103.
19   Gaukroger, *Descartes*, 187–90, examines various reasons for Descartes's retreat from France and explains it in terms of Descartes's lack of a patron and the constraints this placed on his options.

very prominent diplomat and courtier, and the French diplomat Pierre Chanut. Chanut it was who lured Descartes to his death in 1650 (unwittingly, to be sure) by encouraging the Swedish monarch Christina to invite Descartes to Stockholm to tutor her in philosophy. Descartes's keenness to go (he solicited the invitation) indicates his self-perception as the consort of royalty. It should be recalled that one of his greatest works, the *Principles of Philosophy* (1644), had been dedicated to the Princess Elizabeth of Bohemia.[20]

Elizabeth was the eldest daughter of Frederick V, the deposed "Winter King" of the Holy Roman Empire; the family lived in exile in the Netherlands during the period of Descartes's residence there. The philosophical correspondence between Elizabeth and Descartes had begun in 1642, at a time when Descartes had a house a short distance from The Hague, where the princess and other members of her family lived. She began to make a practice of visiting him, sometimes with a party of noble companions who wanted to see the well-known philosopher. Descartes seems to have resented these intruders, and he very soon moved out to a safer distance. From there he could continue to correspond with the princess but to see her only during his own trips to The Hague.[21]

The dedication of the *Principles* was no casual attempt to curry favor. As historians of philosophy well know, throughout the 1640s Descartes and Elizabeth exchange considerable numbers of letters (thirty-three of Descartes's, and twenty-five of Elizabeth's, survive), besides their personal meetings.[22] Elizabeth treated Descartes not only as her tutor and interlocutor in philosophical discussions, but also as her adviser in medical matters. The latter role she construed quite broadly, often seeking Descartes's advice on ways of relieving her frequent bouts of depression brought on by her family's continuing ill fortunes. Descartes's *Treatise on the Passions of th Soul* (1649) was the direct consequence of his advice to Elizabeth.

Descartes's move to the Netherlands, then, cannot be seen, as he himself asserted, as an attempt to keep his head down. He had not moved to a country where he was freed of social obligations, which, indeed, he seems positively to have courted, even to the point, with Elizabeth, of having to resort to minor evasive actions that he could equally well have taken in France. Evidently, Descartes liked living in the Low Countries, just as he liked consorting with the people, of his own or higher social status, that he found there.

Descartes's philosophy had to a considerable degree been shaped during his first Dutch sojourn. As a soldier in 1618, he had fallen in with Isaac Beeckman, a schoolmaster and an enthusiast for something he called

---

20 On Descartes's early acquaintance with Huygens, see ibid., 293; with Elizabeth, ibid., 385–87.

21 Cohen, *Ecrivains français*, 604–7, which uses Samuel Sorbière as its authority about the visits.

22 Most of these letters are translated in Descartes, *The Philosophical Writings*, vol. 3, and in Blom, *Descartes*, with commentary.

"physico-mathematics." In Beeckman's usage, this term designated attempts at explaining physical phenomena by reference to submicroscopic particles and their mechanical interactions, a general picture of the world that was to become the centerpiece of Descartes's mature philosophical thought. And when, two decades later, Descartes began to publish his philosophy with the appearance of the *Discourse on Method*, its most enthusiastic sectaries were Dutch.[23]

At first, Descartes clearly had ambitions to enroll in his support the Jesuits, his own teachers. He wanted his philosophy to take the place of Aristotle's in the standard curriculum of Catholic colleges and universities, and he even dedicated the *Meditations* (1641) to the faculty of the Sorbonne. The Jesuits were a particularly tempting target, owing to their control of a hugely influential network of colleges throughout Catholic Europe, and Descartes made sure that he kept on good terms with them. His lack of success on that front, however, was partially offset by the rapid inroads made by Cartesian philosophy into Dutch universities in the years around 1640, during the heyday of Descartes's publication efforts. The Netherlands were particularly fertile ground for his philosophy, just as they were for the establishment of his persona.

Automata and Morals

The most characteristic component of Descartes's philosophy was his mechanistic picture of the physical world. The plausibility to Descartes of Beeckman's version of this ontology may, it has been suggested, owe much to his encounter, during his soldiering, with the clocks, automata, and other mechanical contrivances that were especially favored in that period by some German princes.[24] The construction of artificial people, animals, and ships run by clockwork had become, by the early seventeenth century, a commonplace of expensive courtly frivolity. The operation of such devices typically accompanied the marking of the hours in the manner of some elaborate late medieval cathedral clocks (of which the Strasbourg example is the most famous). Descartes's writings often

23 Verbeek, *Descartes and the Dutch*; idem, "Regius's *Fundamenta physices*"; Westman, "Huygens and the Problem of Cartesianism."
24 See Mayr, *Authority, Liberty, and Automatic Machinery*, 62–67 on Descartes's mechanistic physiology, chap. 1 on automata an princes; Maurice and Mayr, *The Clockwork Universe*, for many photographs of such devices; Price, "Automata"; Moran, "Prince, Machines." Descartes describes a grotto with automata, similar to one described by Salomon de Caus in a book of 1615, in his posthumously published *Traité de l'homme*, p. 13 of original French edn (1664) reproduced in Descartes, *Treatise of Man*, i.e., Descartes, *Oeuvres*, 11:130; cf. Descartes, *Discours de la méthode*, 420–22. See also Descartes's remarks in his *Principles of Philosophy*, bk. 4, sect. 203; Descartes, *Oeuvres*, 8:326. Rodis-Lewis, *L'oeuvre de Descartes*, 2:469–72, gives much supplementary material.

allude to such devices as a means of elucidating his ideas on the animal (including human) organism.

In the fifth part of the *Discourse on Method*, for example, Descartes discusses the action of the heart, explaining how the blood circulates around the body because of the heart's arrangement of valves and its innate heat. The heat rarefies the blood coming into the heart from the veins, and the valves permit the expanded blood to escape only into the arteries, where it cools and condenses again as it travels along. Descartes clarifies the explanatory virtue of his purely qualitative account in the following terms:

> [T]he movement I have just explained follows from the mere arrangement of the parts of the heart (which can be seen with the naked eye), from the heat in the heart (which can be felt with the fingers), and from the nature of the blood (which can be known through observation). This movement follows just as necessarily as the movement of a clock follows from the force, position, and shape of its counterweights and wheels.[25]

The demonstrative form of the account gives it its credibility, according to Descartes, and his paradigm of intelligibility is a clock.

Otto Mayr has noted that although Descartes frequently drew mechanical analogies in his philosophy of nature, these were almost entirely restricted to the realm of living things – plants and animals, including the human body. The workings of the heavens, for example, a prime target for clockwork metaphors since classical antiquity, received no comparisons with automata of any kind.[26] There is thus a sense in which Descartes's mechanistic universe was at its most authentically mechanical when discussing life, a phenomenon to be elucidated in terms of self-contained, self-moving machines.[27] It should always be remembered that a lawlike, mathematically determinate universe need not be specifically mechanical; Descartes's universe was mechanical only insofar as he used machines, and especially automata, as models of intelligibility.

By midcentury, it was no longer an absurd proposition in France that even human behavior should be intelligible in terms of machines (even if not fully reducible to them). It is well known, for example, that Blaise Pascal spoke of human "machine" behavior (such as often-repeated religious rituals) as a way of establishing beliefs through habit.[28] Louis XIV's Versailles, slightly later in

25   Descartes, *Oeuvres*, 6:50; trans. Descartes, *The Philosophical Writings*, 1:136. See, for a particularly lucid account of Descartes's mechanistic physiology, Hatfield, "Descartes' Physiology"; also Gaukroger, *Descartes*, 269–82.

26   Mayr, *Authority, Liberty, and Automatic Machinery*, 63–64.

27   This point is made in Jaynes, "The Problem of Animate Motion." The classic survey of this issue is Rosenfield, *From Beast-Machine to Man-Machine*. See also Vartanian, "Man-Machine."

28   Pascal, *Oeuvres*, 501–2, from "Pensées," on "La Machine"; see on this Keohane, *Philosophy and the State*, 273.

the century, was sometimes described as a "machine" because of its elaborate courtly ritual and fondness for spectacle that helped constitute the political integrity and power of the king, at the center of the state.[29] But more directly, at the beginning of the century there was already, in the Dutch Republic, what was in some respects a prototypical form of human automation: the organization of its army. The neo-Stoic writings of the Dutch scholar Justus Lipsius had attempted, at the end of the sixteenth century, to renew the virtues of the Roman Empire for the United Provinces, newly emergent from the control of Spain and still fighting for their existence both commercially and militarily. Part of Lipsius's work involved a major treatise, *De militia Romana* (1596), concerning the proper role of armed forces in the constitution of a state, and the manner in which they should be organized; he also wrote at length on Roman military tactics. Lipsius's intention was to promote the Roman model as the right one for the United Provinces. One of the major features of Lipsius's teaching on the military was its stress on the importance of military discipline. This meant the instilling of self-control, restraint, and moderation both in the behavior of individual soldiers and in the collective units made up of them. That was what had made the Roman army so formidable, and should be emulated by the Dutch.[30]

Lipsius was not only a major figure in the intellectual life of this period; his prescriptions, congenial to Prince Maurice of Nassau, were actually put into practice in the remarkably successful armies established by the Dutch. In 1618 Descartes had become a member of one of the two French regiments in Maurice's standing army, and would have seen at first hand the disciplined ethic by which it was governed. Descartes, at that point in his life, was exposed not only to the literal automata beloved of German princes, but also the human automata being drilled and disciplined in the Dutch military encampments.[31]

29   This is the central theme of Apostolidès, *Le roi-machine*; see also Revel, "La cour."
30   On neo-Stoicism in this period see Oestreich, *Neostoicism*; on Lipsius, with especial focus on his neo-Stoic natural philosophy, Saunders, *Justus Lipsius*. There is a growing body of literature on Stoic natural philosophy in this period: Barker and Goldstein, "Is Seventeenth-Century Physics Indebted to the Stoics?"; Barker, "Jean Pena"; idem, "Stoic Contributions"; Freudenthal, "Clandestine Stoic Concepts."
31   Oestreich, *Neostoicism*, chap. 5 (on military reform in the Netherlands); see also Gaukroger, *Descartes*, 65–67 on Descartes's military experience. An insightful treatment of Maurice's innovations in regard to drill and its concomitant disciplinary as well as technical efficacy may be found in McNeill, *The Pursuit of Power*, 125–39, which also notes the spread of the new training regime to other European armies in emulation of the Dutch; also Parker, *The Military Revolution*, 19–22 (more generally on communal bodily discipline, see also McNeill, *Keeping Together in Time*). Simon Schaffer has drawn my attention to Franz Borkenau's identification, in the 1930s, of Descartes's philosophy with Dutch neo-Stoicism. Borkenau, like Weber, associates the latter with a Calvinist theological outlook (as an expression of bourgeois ideology) with which he then associates Descartes: "coming from the gentry, [Descartes] builds on the presupposition of stoic morality." Borkenau, "The Sociology of the Mechanistic World-Picture," 120. Borkenau links

It is thus not surprising to find that the highest aim of Cartesian philosophy was proper behavior. In his preface to the French translation of the *Principles of Philosophy* in 1647, Descartes characterizes the philosophy of which his treatise speaks as a tree. "The roots are metaphysics, the trunk is physics, and the branches emerging from the trunk are all the other sciences."[32] Physics, of the kind discussed in the *Principles*, thus gives forth all the special sciences, which have (in the last analysis) to be understood in its terms. Furthermore, these latter "may be reduced to three principal ones, namely medicine, mechanics and morals [*morale*]. By 'morals' I understand the highest and most perfect moral system, which presupposes a complete knowledge of the other sciences and is the ultimate level of wisdom."[33] The study of philosophy, he had observed a little earlier, "is more necessary for the regulation of our morals and our conduct in this life than is the use of our eyes to guide our steps."[34]

"Medicine, mechanics and morals" make a strange triptych nowadays, perhaps, but they made sense to Descartes. The first, medicine, was a lifelong concern of his, one that often provided him with an account of the chief benefit to be derived from a true physics. The prolongation of human life, to a practically preternatural extent, is a recurring theme in his writings, especially in the correspondence, and it is therefore no surprise that Descartes insisted on doctoring himself. It evidently came as a shock to his friends to learn of his demise, in Sweden, at the unduly modest age of fifty-four. Queen Christina sneered that "ses oracles l'ont bien trompé." (One of those oracles was evidently constituted by his teeth, much like a horse; in 1639 he reckoned that their good condition indicated that he had another good thirty years at least, barring accidents.) The red wine infused with tobacco that he insisted on treating himself with during his final illness brought up the phlegm but brought down the philosopher.[35]

Descartes's insistence on visualizability in his natural philosophy and mathematics (especially his insistence on geometrical representations in mathematics) to "the handicraft basis of production" in this period (ibid., 121); cf. my argument regarding automata in this section. Another Marxist treatment of such issues is Zur Lippe, *Naturbeherrschung am Menschen*; see esp. vol. 2. It is, of course, important not to overemphasize a supposed dominant bourgeois morality in the Netherlands at this time: for a less idealized picture, see Schama, *The Embarrassment of Riches*, chap. 3.

32   Descartes, *Principes*, author's letter, in Descartes, *Oeuvres*, 9:14 (*Principes* has its own, separate pagination within the volume); trans. Descartes, *The Philosophical Writings*, 1:186.

33   Ibid. Note that these three are all comprised under the general heading of "physics" (a contemporary synonym for "natural philosophy"), which would not traditionally have included ethics. Evidently Descartes is intent on establishing "physics" as covering the entire realm of creation.

34   Descartes, *Oeuvres*, 9:3–4; trans. Descartes, *The Philosophical Writings*, 1:180.

35   See the presentation of relevant materials in Lindeboom, *Descartes and Medicine*, esp. 94 on Christina and teeth; Gaukroger, *Descartes*, 416, discusses Descartes's death.

In the preface to the French *Principles*, Descartes had also proposed that "a nation's civilization and refinement depends on the superiority of the philosophy which is practised there."[36] According to Norbert Elias, this notion of "civilization" arose in France and elsewhere in the early modern period as an aspect of court society and the associated social stratifications that had begun to coalesce in the sixteenth century.[37] Elias finds the French word *civilisation* in use no earlier than the middle of the eighteenth century,[38] but the concept of *civilité*, closely associated with allied terms such as *politesse* and *gentillesse*, or with the ideal of the *honnête homme*, was in common currency to denote a condition both of individuals and of societies. It is not surprising that Descartes, an *honnête homme* born and bred, and an assiduous philosophical courtier, should identify with it.

Good philosophy leads to good behavior, and good behavior is "civilized" behavior. The considerable courtesy literature of the sixteenth and seventeenth centuries, notably of French, Italian, and English provenance, details the specifics of this kind of good manners; its relation to Descartes's philosophy seems, on the face of it, less evident. How could mechanistic philosophy and the "method" lead to civilization and refinement? There can be no doubt, as we have already seen, of Descartes's concern with these latter attributes. But even the *Discourse on Method* displays the attributes expected of an *honnête homme*. [. . .]

[. . .] The readers of courtesy literature sought counsel on how to conform to an alien morality. Their fear of missteps was a fear of revealing to competent members of court society that they did not possess full membership in that moral community. But that moral community itself had pretensions to an inwardness that was merely expressed in the outward forms of manners. Descartes's various remarks on *morale* (culminating in his *Treatise on the Passions of the Soul* of 1649) show, as do other such contemporary writings, the deep seriousness of literature on manners.

In a world where behavior was so powerfully controlled by one's place in a social system of remarkably precise, and obvious, ordering, what seemed appropriate in social interaction and what seemed appropriate in the dispositions of the natural world – that is, what was regarded as "natural" in each – had certain points of contact that bound them together.[39] Cartesianism in the

---

36   Descartes, *Oeuvres*, 9:3; trans. Descartes, *The Philosophical Writings*, 1:80. Descartes talks of a nation "plus civilisée & polie."

37   Elias, *The Civilizing Process*, "The History of Manners," part 1, chap. 1; part 2.

38   Ibid., 241 n. 25. *The Oxford English Dictionary* similarly gives no examples of the English word *civilization* prior to the eighteenth century. For additional references on the provenance of the word, see Fox, "Introduction," 29 n. 111.

39   It ought to be observed at this point that the structure of the present argument is not in the mold of many of the contributions to that hoary classic, Barnes and Shapin, *Natural Order*. Some of those essays set up isomorphisms between ideas about an aspect of the natural world and

Netherlands owed its success and its basic point precisely to new forms of personal behavior and their associated sensibilities. The meaning of Descartes's mechanical universe resided in its ability to make sense of those new forms. Thus Dutch society was especially ready to embrace Descartes's work because Cartesianism was a natural philosophy for a bourgeois society.

Descartes's way of formalizing the idea and scope of a machinelike component of human behavior involved restructuring the established concept of "souls." The part-theological, part-psychological genre into which Descartes's writings on the soul and the passions fall has been examined by Nannerl Keohane.[40] She identifies a number of important themes that reappear in seventeenth-century French discussions of love, a principal concern in treatments of the passions. An Augustinian conceptualization predominated: it involved a distinction between two kinds of love, the one pure, selfless, and directed toward God, the other self-interested – *amour-propre*. The *littérateur* Guez de Balzac, a correspondent of Descartes, emphasized the Augustinian distrust of self-love and advocated participation in the wider community as the proper condition of humanity: "Each individual is not enough even to be one unless he tries to multiply himself in certain ways with the help of many; and to consider us in general, it seems that we are not so much whole bodies as disconnected parts that are reunited in society."[41] Descartes expressed similar sentiments in a letter of 1645 to Elizabeth, as also in *Passions of the Soul*.[42] In the wider context of an ongoing discussion of Seneca, Descartes observes that "one could not subsist alone and is, indeed, one of the parts of the earth, and more particularly, of this state, of this society, of this family, to which one is joined by one's residence, one's fealty, and one's birth."[43]

Descartes's varied strategies in his writings depended to a large extent on the cultural audience to which he wished to make appeal. In France, which remained the reference point of his intellectual world, the competing principles of the elites of the sword and the robe, of the erudite, humanist court speaking French and the academic, disputatious university speaking Latin, required different modes of presentation. Both were adopted by the Jesuits, but they were clearly distinct in style.[44] Descartes's switching between French and

attitudes toward the social order held by the same people, and infer a causal link from the former to the latter (e.g. Wynne), while others take a practically instrumental view of the function of particular ideas about nature for the furthering of their proponents' social interests (e.g. Barnes and MacKenzie). I shall, by contrast, be attempting to show that appropriate social behavior and appropriate behavior toward nature were fundamentally the same thing for Descartes and many of his contemporaries.

40 Keohane, *Philosophy and the State*, chap. 6; see also Levi, *French Moralists*, chap. 10.
41 Quoted in Keohane, *Philosophy and the State*, 200 (from Balzac's *Aristippe*).
42 See the discussion in ibid., 204–8.
43 Descartes to Elizabeth, 15 September 1645, in Descartes, *Oeuvres*, 4:290–96; cf. Blom, *Descartes*, 151 (my translation deviates from Blom's).
44 Fumaroli, *L'âge de l'éloquence*, esp. 247–56.

Latin for his publications, and the increasing formalism of the *Principles of Philosophy* and *Meditations*, with their original Latin versions in the 1640s, as contrasted with the easy style of the French *Discourse on Method* of a few years earlier, seems to mirror that cultural tension. The humanist stress on the importance of rhetoric and the centrality of imitation in rhetorical pedagogy dominated Jesuit education, however, as did the basic assumption that, as Juan Luis Vives had observed in the sixteenth century, "a true imitation of what is admirable is a proof of the goodness of the natural disposition."[45] Descartes's philosophic project furthered the humanist assumptions of the court much as did Jesuit pedagogy, even while maintaining a link between the two cultures.

The portions of Descartes's projected natural philosophy that were to deal with the nature and relationship of the human body and soul had been outlined in the *Discourse*. Descartes seems never to have finished his intended account (at least, it does not survive), although the *Passions of the Soul* goes some way toward filling the gap. But in objecting to the views of his erstwhile disciple, the Dutchman Regius, in 1641, Descartes insisted that there is only one kind of soul, the human rational soul. All other vegetative and animal properties are due to the arrangement of bodily parts.[46] Furthermore, the synopsis in the *Discourse* presents Descartes's position quite clearly. Perhaps the most telling passage concerns the aforementioned question of the differences between automata and humans. He wants to stress not only that the living human body, like other animal bodies, is a kind of elaborate automaton of the sort that God would be capable of fabricating but also that human beings are not *just* automata. He says:

> if any such machines had the organs and outward shape of a monkey, or of some other animal that lacks reason, we should have no means of knowing that they did not possess entirely the same nature as these animals; whereas if any such machines bore a resemblance to our bodies, and imitated our actions as closely as possible for all practical purposes, we should still have two very certain means of recognizing that they were not real men.[47]

The first way lies in seeing that they were unable to use language. Descartes notes that an automaton could be made to pronounce words,[48] and even to

45 Vives, *Vives: On Education*, 194.
46 Consecutive letters to Regius, May 1641, in Descartes, *Oeuvres*, 3:369–70, 370–72 (cf. Descartes, *Treatise of Man*, 114). On souls in Renaissance philosophy, see Park, "The Organic Soul." For an argument stressing the role of more general "philosophical" reasons in Descartes's rejection of a vitalist conception of life, rather than reasons derived from within the anatomical tradition and based on more empirical grounds of practical intelligibility, see Sloan, "Descartes, the Sceptics, and the Rejection of Vitalism."
47 Descartes, *Oeuvres*, 6:56; trans. Descartes, *The Philosophical Writings*, 1:139–40.
48 Bedini, "The Role of Automata," discusses the long tradition of talking statues that this example invokes.

respond to certain stimuli, as, for example, "if you touch it in one spot it asks you what you want of it"; nonetheless, he is convinced that "it is not conceivable that such a machine should produce different arrangements of words so as to give an appropriately meaningful answer to whatever is said in its presence, as the dullest of men can do."[49] Presumably the difficulty would lie in the sheer number of possible conversations in which the automaton might be required to participate.[50] One might, of course, object that if we imagine God as the automaton's artificer, His omnipotence would allow any number of diverse stimulated linguistic responses to be "programmed" into the machine. Descartes's second means of identifying the nonhuman impostor, however, may be seen as addressing that difficulty.

> Secondly, even though such machines might do some things as well as we do them, or perhaps even better, they would inevitably fail in others, which would reveal that they were acting not through understanding but only from the disposition of their organs. For whereas reason is a universal instrument which can be used in all kinds of situations, these organs need some particular disposition for each particular action; hence it is for all practical purposes impossible for a machine to have enough different organs to make it act in all the contingencies of life in the way in which our reason makes us act.[51]

Descartes's notion of the equality or even superiority of machines to humans in particular tasks resembles the sociologist Harry Collins's idea of "behavior-specific action," or "machine-like action." This is a kind of action the description of which is entirely exhausted by a specification of its characteristic behavioral coordinates – the physical motions in space and time. Machines can in principle emulate this kind of action perfectly, but they cannot adapt to new circumstances in which the "same" action might now correspond to a different set of behavioral coordinates. So too with Descartes's automata.

Descartes's own explanation of the distinction is expressed in terms of the capacity of human beings to use reason, the expression of their immaterial *res cogitans*. Reason allows a person to adapt to changing circumstances – "all the occurrences of life" – in appropriate ways. By contrast, Collins makes his demarcation on an importantly different level. Rather than speaking of "reason," a faculty possessed by individuals acting so as to produce behavior appropriate to the situation, Collins highlights social life, by definition shared

---

49   Descartes, *Oeuvres*, 6:56–57; trans. Descartes, *The Philosophical Writings*, 1:140.
50   This argument is very similar to the (much more elaborate) one given in Collins, *Artificial Experts*, chap. 14: see below.
51   Descartes, *Oeuvres*, 6:57; trans. Descartes, *The Philosophical Writings*, 1:140. See also Descartes, *Discours de la méthode*, 423–25, for further material from Descartes's correspondence reiterating the same point at greater length, and further references.

among individuals. It is that social life which creates the ever-shifting context within which appropriate action is created and validated. A machine cannot take part in human social life, according to Collins, and hence is excluded from the possibility of genuinely human action – "acting through understanding," in Descartes's phrase. Descartes's talk of "understanding," or "reason," is an expression of Collins's Wittgensteinian notion of "form of life," and it is fundamentally social.

Passions

In his *Treatise on the Passions of the Soul* in 1649, Descartes relates his human-izing category of "reason" to internal emotional states by explaining these states mechanistically. The little book had originally been written for the benefit of Princess Elizabeth, who apparently suffered from bouts of depression brought on by her family's ill-fortune. As many commentators have observed, Descartes's treatise fits squarely within a genre of writing on the passions that had both classical antecedents (notably Epicurean and Stoic) and modern exemplars, often directly integrated with theological issues.[52] The closest precedents to Descartes's own approach were Stoic, again paralleling the neo-Stoicism of Justus Lipsius, and like Lipsius, Descartes stressed the goal of self-mastery.

The "passions" are disturbances that affect the mind; they are so called because the mind is affected by them as a patient, not produced by it as an agent. Specifically, Descartes is concerned with the kind of passions that, while often triggered by physiological events, are not experienced as external senti-ments, as of pain in a limb or the heat of a fire, but as purely internal, emo-tional states, such as sadness or joy.[53] The passions affect the mind by means of a naturally established relation between their particular physical manifes-tations and the character of the mind's apprehension of them through their effect on the flow of (entirely material) "animal spirits" through the pineal gland in the brain – so that the dryness of the throat will serve to produce an active desire to drink, for example. But they can also themselves be affected by the mind, again by the redirection of spirits through the pineal gland. The indi-viduality of Descartes's book stems in large part from its stress on the mind's capacity for controlling the passions to its own advantage. The mind, in this

52   See Descartes, *Lea passions de l'âme*, "Introduction," esp. 21–32; Rodis-Lewis, introduction to Descartes, *The Passions of the Soul*; and refs. in n. 46, above. Gaukroger, *Descartes*, clasp. 10, is an exemplary discussion of the *Passions*, both the arguments therein and the context for the text's production.

53   Descartes identifies three different kinds of passions strictly speaking; only the kind in which the mind is described as acting on itself is of relevance in his discussions of passions qua inner emotional states on urges: Descartes, *Passions*, arts. 17–27.

view, is not to be trained merely to restrain the tendency of the passions to cloud judgment; it is also to be trained to use the passions actively for positive ends. The otherwise rather similar Stoic doctrine sought only the first goal, that of subduing the passions.[54] Thus pleasure can be gained from the proper handling of the passions (Descartes instances theatrical performances to show that even apparently unpleasant passions such as fear or sadness can sometimes engender pleasure), or courage may be summoned up when needed.[55] And while Juan Luis Vives, whose own discussion of the passions Descartes cites, had couched his account of the physiological disturbances that correspond to each passion in terms of orthodox Galenic humoral theory, Descartes describes them in mechanistic terms.[56] The mind controls the passions, without disdaining them, through the use of reason and the will to control the machine of the body.

Descartes's understanding of the essence of human beings was inseparable from his perception of machines. Since Descartes regarded the human body as a kind of machine, so that the action of the heart, as well as all other organs and parts, was subject to the same mechanical necessity as that of a clock, it might appear that a mechanical physics of the kind outlined in the *Principles* could have no connection with manners and morality – which inhabited a realm categorically different from that concerning the physical springs of bodily behavior. But Descartes's account of the passions of the soul shows how the mechanistic human body, described in greatest detail in the *Traité de l'homme* (ca. 1633) and the brief *Description du corps humain* (late 1640s),[57] could interact with the mind so as to produce behavior that was an amalgam of the two substances. That behavior was therefore neither wholly mechanical nor wholly rational; morality consisted of an appropriate accommodation of the two. In the *Description*, Descartes begins by saying: "There is no more fruitful exercise than attempting to know ourselves. The benefits we may expect from such knowledge not only relate to ethics, as many would initially suppose, but also have a special importance for medicine."[58]

The self to be understood through such means was the self that Descartes described in the *Passions*. The body acts on the mind through the passions, and the mind reacts on the passions by its control of the body through the pineal gland. Unlike animals, which lack incorporeal, rational souls, human beings can therefore endeavor to control even those parts of their behavior that are

54   On the general, and considerable, neo-Stoic elements of Descartes's moral views, see esp. Levi, *French Moralists*, esp. 241–48 (up to and including the *Discourse*), and chap. 10 on the *Passions* and attendant correspondence with Elizabeth, which provides further references. On neo-Stoicism in this period, see above, n. 30.
55   Descartes, *Les passions de l'âme*, part 3, "Des Passions particulieres."
56   Vives, *The Passions of the Soul*; see also Noreña, *Juan Luis Vives*.
57   Both unpublished in his lifetime.
58   Descartes, *Oeuvres*, 11:223; trans. Descartes, *The Philosophical Writings*, 1:314.

directly caused by the actions of the body.[59] As Vance G. Morgan usefully puts it, the passions "are a unique manifestation of the mind-body union in the human being, and their primary purpose is to aid in the preservation of that union"[60] – by naturally inclining us toward actions that will usually be beneficial.[61] Many of the passions, however, represent involuntary physical expressions of mental states (such as joy, desire, hate, and so forth). Mastering the passions therefore requires self-discipline.

> For anyone who has lived in such a way that his conscience cannot reproach him for ever having failed to do anything he judged to be best (which is what I call following virtue here) derives a satisfaction with such power to make him happy that the most vigorous assaults of the Passions never have enough power to disturb the tranquillity of his soul.[62]

Gestures and facial behavior were classic expressions of the passions and accordingly receive much treatment in Descartes's treatise. Thus people flush, tremble, cry, turn pale, laugh, weep, sigh, and in many other ways seem to betray their inner feelings.[63] In particular, "there is no Passion which is not manifested by some particular action of the eyes."[64] Descartes goes on to discuss the subtleties of expression discernible in the "movement and shape of the eye" as well as in facial behavior generally and notes the frequent difficulty found in making sharp distinctions between them. "It is true that there are some that are quite recognizable, like a wrinkled forehead in anger and certain movements of the nose and lips in indignation and mockery, but they do not seem to be natural so much as voluntary."[65] Descartes now makes the bridge between physical expressions of the passions and the place of such expressions in social life: "And in general all the actions of both the face and the eyes can be changed by the soul, when, willing to conceal its passion, it forcefully imagines one in opposition to it; thus one can use them to dissimulate one's passions as well as to manifest them."[66] Not only can the inward manifestation of

59   Cf. Descartes, *Les passions de l'âme*, art. 138.
60   Morgan, *Foundations of Cartesian Ethics*, 165. Ibid., chap. 5 is a particularly clear account of Descartes on the passions; see also Rorty, "Cartesian Passions."
61   See the Sixth Meditation (Descartes, *Oeuvres*, 7:84–88) for Descartes's explanation of how this relation can sometimes go awry.
62   Descartes, *Les passions de l'âme*, art. 148, trans. in Descartes, *The Passions of the Soul*.
63   See in particular ibid., arts. 114–36. For the role that music was taken to play in inciting the passions, and the operational control that music theorists such as Marin Mersenne thought might thereby be attained over the passions, see Duncan, "Persuading the Affections"; Montagu, *The Expression of the Passions*, 55.
64   Descartes *Les passions de l'âme*, art. 113 (trans. Voss).
65   Ibid.
66   Ibid. The edition by Stephen Voss (Descartes, *The Passions of the Soul*) includes plates from Charles Le Brun, *Conférence sur l'expression générale et particulière* (Paris, 1696), which depict (for artists) typical facial expressions corresponding to a conventional set of passions. See Ross, "Paint-

the passions be controlled (their "assault" on the soul), but their outward manifestation as well. Such issues of gesture and facial behavior were important matters in seventeenth-century social interaction; Elias and others have documented the self-conscious concern with which people handled them.[67] The courtesy and civility literature of the period discusses at great length these issues and the importance of their mastery for social interaction.

Descartes's account of the passions exploits a rigid distinction between them and reason which is rooted in a mechanistic ontology effectively *defined* by the mind-body distinction. It is this distinction, and this ontology, which allows the individual to make his or her own persona. Elizabeth, whose woes and unhappiness largely prompted Descartes to expatiate on these matters at such length, owed her difficulties to a social standing the ordinary expectations of which were thwarted – she was a princess in exile. A milkmaid would not have grieved overmuch for want of a kingdom, but Elizabeth did. Descartes comforted her by advising a moral stance that portrayed her troubles as manageable and her determined behavior as consonant with her status.

It is noteworthy that Descartes does not discuss sexual differences when giving his physiological accounts of the passions. At a less explicit level, one might expect gender differences to appear in the presentation of examples, but these are determinedly rooted at the level of generic human behavior regarding the passions themselves, gender playing only an incidental modulating role.[68] It is this relative gender neutrality at the physiological and mental, if not social, levels in Descartes's philosophy that led some philosophical women in the seventeenth century to adopt Cartesianism as a badge of emancipation.[69]

ing the Passions"; Montagu, *The Expression of the Passions*, esp. chap. 1 on Le Brun's theory and its avowed indebtedness to Descartes.

67   Elias, *The Civilizing Process*, and refs. in n. 41, above; Bremmer and Roodenburg, esp. articles by Burke, "The Language of Gesture," Roodenburg, "The 'Hand of Friendship,' " and Muchembled, "The Order of Gestures."

68   As, for example, Descartes, *Les passions de l'âme*, arts. 82, 168, referring to the passions of a brutish man for the woman he wants to rape and to the "honorable" passion of a woman to protect herself from such treatment. These examples tend to escape gendered specificity in regard to the accounting of particular passions; such cases are illustrated by a variety of examples besides those just given, and none is identified with women rather than men. Thus *Passions*, art. 168, on "honorable passions," is also illustrated by examples in which men are the subjects.

69   See Harth, *Cartesian Women*. Harth (67–78) discusses the dedicatee and prime inspirator of the *Passions*, Elizabeth, noting (75–76) Elizabeth's ironic greater readiness than that essayed by Descartes to invoke her physiological characteristics as a woman in discussing the management of her melancholy passions. Gaukroger, *Descartes*, 468 n. 93, comments on art. 147 of *Passions*, regarding the case of a man outwardly grief stricken but secretly glad about the death of his wife, that it shows a "low view of women" by Descartes insofar as it does not seem plausible that Descartes might have presented the story the other way around, with the woman secretly glad of her husband's death. However, this case could just as easily be read as critical of the grieving husband for hypocrisy.

Thus, for a mechanist like Descartes, and for the legion of his followers that sprang up in the 1630s and '40s (including the young Christiaan Huygens), nature was made intelligible through the idea of machinelike action. And at the heart of Descartes's philosophical project lay the assumption that this kind of intelligibility also serves to make people intelligible: the behavior of people and the behavior of machines are in large measure semantically identical. The residual differences between them that the Cartesian philosophy maps out are to be accounted for in terms of the distinction between the behavior of machines and the behavior of reasoning agents, a distinction that helps to define what machines and reason really are. In a strong sense, as I have argued, Cartesian "reason" is a form of socialization. [. . .]

Freedom and Autonomy

In the *Principles of Philosophy*, there is a section headed: "The supreme perfection of man is that he acts freely or voluntarily, and it is this which makes him deserve praise or blame."[70] This freedom was precisely what machines lacked.

> The extremely broad scope of the will is part of its very nature. And it is a supreme perfection in man that he acts voluntarily, that is, freely; this makes him in a special way the author of his actions and deserving of praise for what he does. We do not praise automatons for accurately producing all the movements they were designed to perform, because the production of these movements occurs necessarily. It is the designer who is praised for constructing such carefully-made devices; for in constructing them he acted not out of necessity but freely. By the same principle, when we embrace the truth, our doing so voluntarily is much more to our credit than would be the case if we could not do otherwise.[71]

Praise and blame attach only to the actions of a free agent, and machines are not free. Such a point would not need to be made if matters were not already being represented in ways that made the opposite look increasingly plausible. Automatic machines, one might say, represent a model of predestination where the omniscience of the clock maker stands in for that of God. But the metaphor (if that is all it is) also abandons explicitly any semblance of free will, the chief function of Descartes's separation of the mind from the body.

As we saw above, Descartes counted judgment and will as the two primary properties of the mind, the *res cogitans*. "Judgment" is a faculty for using

---

70 *Principia*, bk. 1, sect. 37: Descartes, *Oeuvres*, 8A:18; trans. Descartes, *The Philosophical Writings*, 1:205.
71 Ibid.

reason, typically on material provided from outside; the will, however, is the faculty that renders the human soul truly free.[72] There are traces here of Averroism, wherein the "active" soul, which instantiates reason, is unitary and unchanging; everyone participates in it since reason is the same for all. Human psychic individuation, according to Averroës, results from the "passive" soul, which is shaped by the individual's unique experiences.[73] Since Descartes often speaks of the will, or volition, as the central autonomous characteristic of the soul, rather than of judgment or reason, one can see the will for Descartes as serving the same role as did the passive intellect for Averroës; the freedom of the individual will is the only identifying mark of nonmaterial human individuality, and the extent of its scope is therefore all-important.

It is therefore unsurprising that Thomas Hobbes, a more uncompromising materialist than Descartes or any of his other critics, took issue with Descartes in the Third Objections to the *Meditations* on just this point. Hobbes, regarding Descartes's discussion of the possibility of error in human knowledge, complains that it should be noted "that the freedom of the will is assumed without proof, and in opposition to the view of the Calvinists."[74] Hobbes was all in favor of causal determinism (of a specifiably mechanical sort) as the primary criterion of intelligibility in philosophy, and he attempts to buttress his position by equating it with Calvinist theology. Hobbes wants Calvinist predestination to be fully instantiated by automata, whereas Descartes wants to leave an escape route;[75] the lack of freedom exhibited by machines exempts them from moral judgments that are unquestionably appropriate for human beings.

However, while final causes in the Aristotelian sense are noticeably absent from Descartes's construal of his machine paradigm, they cannot be said to be absent from the paradigm itself: the machine metaphor does not determine the nature of machines. Clocks have a function, but Descartes does not regard that function as descriptive of what they are in themselves *as machines*: like Boyle, he would have thought that he "had fairly accounted for it, if, by the shape, size, motion &c. of the spring-wheels, balance, and other parts of the watch [he] had shown, that an engine of such a structure would necessarily mark the hours."[76] Descartes's vision of natural philosophy differs signally from that

72   See esp. the Fourth Meditation, Descartes, *Oeuvres*, 7:58–62, and discussion in Morgan, *Foundations of Cartesian Ethics*, 144–45.
73   Kessler, "The Intellective Soul," for the fortunes of this doctrine in the Renaissance. Gaukroger, *Descartes*, 391–92, makes a similar observation regarding Descartes's remarks on "intellectual memory" and its restriction to universals.
74   Descartes, *Oeuvres*, 7:190; Descartes, *The Philosophical Writings*, 2:133.
75   For more on Hobbes's materialism and its implications, see Sarasohn, "Motion and Morality."
76   Boyle, *Hydrostatical Discourses*, quoted in Shapin and Schaffer, *Leviathan and the Air-Pump*, 216. The present point is discussed at greater length in Dear, *Discipline and Experience*, chap. 6, sect. 1.

of scholastic orthodoxy. In the *Principles of Philosophy* Descartes argues that "[i]t is not the final but the efficient causes of created things that we must inquire into."[77] He explains that "[w]hen dealing with natural things we will, then, never derive any explanations from the purposes which God or nature may have had in view when creating them and we shall entirely banish from our philosophy the search for final causes."[78] This is because

> we should not be so arrogant as to suppose that we can share in God's plans. We should, instead, consider him as the efficient cause of all things; and starting from the divine attributes which by God's will we have some knowledge of, we shall see, with the aid of our God-given natural light, what conclusions should be drawn concerning those effects which are apparent to our senses.[79]

We can only understand what God has made, not why He has made it. And the criteria of intelligibility to be applied to this task are rooted in the perceived intelligibility of automata. That is the fundamental point of Descartes's mechanical philosophy. Automata, most familiarly clocks, came to exemplify the transparency of nature to the understanding. And yet they were themselves artificial contrivances.[80]

Descartes's kind of natural knowledge had thus become, rather than an identification of purposes in nature, an attempt at characterizing the "rules" of nature – or what in the seventeenth century are increasingly called *laws* of nature.[81] The rules governing nature's behavior take the place of the purposes for the sake of which that behavior occurs: we want to know how to handle a sword, not why we are fighting. Indeed, fencing is swordplay without instrumental purpose, just as dancing is perambulation without a destination or dressage horsemanship without a battle to win; it is just what a gentleman does.

The amorphousness of our view of the seventeenth century's philosophical mechanism is due to a failure to examine the contemporary meaning of contrivance itself. The distinction between art and nature was dissolved by *fiat*:

77   *Principia*, bk. 1, sect. 28: Descartes, *Oeuvres*, 8A:15; Descartes, *The Philosophical Writings*, 1:202. On final causes and Descartes's mechanical explanations, Rodis-Lewis, "Limitations of the Mechanical Model"; see also Osler, *Divine Will and the Mechanical Philosophy*, 212–13, noting some of the subtleties of Descartes's attitudes toward final causes; also Garber, *Descarles' Metaphysical Physics*, 273–74.
78   *Principia*, bk. 1, sect. 28: Descartes, *Oeuvres*, 8A:15; Descartes, *The Philosophical Writings*, 1:202; the final clause is an addition found in the French *Principes* (Descartes, *Oeuvres*, vol. 9).
79   *Principia*, bk. 1, sect. 28: Descartes, *Oeuvres*, 8A:15–16; Descartes, *The Philosophical Writings*, 1:202.
80   Rossi, "Hermeticism," esp. 252–53, considers the mechanistic view of the world as a criterion of intelligibility.
81   Milton, "The Origin and Development"; idem, "Laws of Nature"; Oakley, *Omnipotence, Covenant, and Order*, 77–92; Zilsel, "The Genesis of the Concept of Physical Law"; also Funkenstein, *Theology and the Scientific Imagination*, 192–93; Ruby, "The Origins of Scientific 'Law,'" on medieval precedents.

the possibility of saying that art is a matter of manipulating rather than over-riding nature relied on shifting the focus from ends to means. Things in the world became made things, and specifying the agency that had made them revealed no essential difference. The import of mechanism was therefore at root methodological rather than ontological. Descartes presented this idea through ontological talk, but he made it work through contrivance. In the end, God could not tell human beings what His purposes were, and human beings could not tell each other what their own purposes were.

Charles Le Brun, the midcentury artist and writer on facial expressions as manifestations of the passions, claimed to derive the physiological warrant for his teachings from Descartes's *Treatise*. He also produced many of the paint-ings that decorated the palace at Versailles, as well as producing the design for the (imperfectly realized) gardens. Sébastien Le Clerc, the geometrician who had used the gentlemanly art of fencing to illustrate his lessons, was himself responsible for the plates that illustrated Charles Perrault's 1677 *Le labyrinthe de Versailles*.[82] Le Brun, like Le Clerc, surely knew the significance of courtli-ness and the meaning of the outward signs of kingly authority, just as he claimed to know the meaning of the outward signs of inner emotion and their propriety.[83] Social life, formalized through manners, meant letting people see each other as automata under the control of reason; so automata is what they became.[84] Alongside the discipline of the Dutch, we are left with the image of Louis XIV amusing himself, toward the close of the century, by spending hours playing with a small automation.[85]

Acknowledgments

I thank Jacques Revel, Simon Schaffer, Harry Collins, and Hal Cook for their valuable comments on earlier versions of this chapter.

References

Apostolidès, Jean-Marie. *Le roi-machine: Spectacle et politique au temps de Louis XIV.* Paris: Editions de Minuit, 1981.
Baillet, Adrien. *Vie de Monsieur Descartes.* 2 vols. Paris, 1691.
Barker, Peter. "Jean Pena and Stoic Physics in the 16th Century." In Ronald H. Epp, ed., *Spindel Conference 1984: Recovering the Stoics (Southern Journal of Philosophy*, supple-ment). 18 (1985): 93–108. (Supplement has separate pagination.)

82   Montagu, *The Expression of the Passions*, 43–45; Perrault, *Le labyrinthe de Versailles*.
83   Apostolidès, *Le roi-machine*, esp. 86–92.
84   Porter, *Trust in Numbers*, esp. chap. 4, discusses the analogous sense in which the cultural dominance of statistics in public affairs tends to remake a society in its own image.
85   Apostolidès, *Le roi-machine*, 138.

——. "Stoic Contributions to Early Modern Science." In *Atoms, Pneuma, and Tranquillity: Epicurean and Stoic Themes in European Thought*, ed. Margaret J. Osler, 135–54. Cambridge: Cambridge University Press, 1991.

Barker, Peter, and Bernard R. Goldstein. "Is Seventeenth-Century Physics Indebted to the Stoics?" *Centaurus* 27 (1984): 148–64.

Barnes, Barry, and Steven Shapin, eds. *Natural Order: Historical Studies of Scientific Culture*. Beverly Hills, Calif.: Sage, 1979.

Bedini, Silvio A. "The Role of Automata in the History of Technology." *Technology and Culture* 5 (1964): 24–42.

Blom, John J. *Descartes: His Moral Philosophy and Psychology*. New York: New York University Press, 1978.

Borkenau, Franz. "The Sociology of the Mechanistic World-Picture." *Science in Context* 1 (1988): 109–27.

Bremmer, Jan, and Herman Roodenburg, eds. *A Cultural History of Gesture*. Ithaca: Cornell University Press, 1991.

Bryson, Anna. "The Rhetoric of Status: Gesture, Demeanour and the Image of the Gentleman in Sixteenth- and Seventeenth-Century England." In *Renaissance Bodies: The Human Figure in Renaissance Culture c. 1540–1660*, ed. Lucy Gent and Nigel Llewellyn, 136–53. London: Reaktion Books, 1990.

Burke, Peter. "The Language of Gesture in Early Modern Italy." In Bremmer and Roodenburg, 71–83.

Camporesi, Piero. *The Anatomy of the Senses: Natural Symbols in Medieval and Early Modern Italy*. Trans. Allan Cameron. Cambridge: Polity Press, 1994.

Cervantes, Miguel. *El licenciado vidriera*. In *The Portable Cervantes*, trans. Samuel Putnam, 760–96. New York: Viking, 1951.

Cohen, Gustave. *Ecrivains français en Hollande dans la première moitié du XVIIe siècle*. Paris: Éduard Champion, 1920.

Collins, H. M. *Artificial Experts: Social Knowledge and Intelligent Machines*. Cambridge: MIT Press, 1990.

Cottingham, John, ed. *The Cambridge Companion to Descartes*. Cambridge: Cambridge University Press, 1992.

Dainville, François de. "L'enseignement des mathématiques dans les Collèges Jésuites de France du XVIe au XVIIe siècle." *Revue d'histoire des sciences* 7 (1954): 6–21, 109–23.

——. *La géographie des humanistes*. Paris: Beauchesne, 1940.

Dear, Peter. *Discipline and Experience: The Mathematical Way in the Scientific Revolution*. Chicago: University of Chicago Press, 1995.

——. "Mersenne's Suggestion: Cartesian Meditation and the Mathematical Model of Knowledge in the Seventeenth Century." In *Descartes and His Contemporaries*, ed. Roger Ariew and Marjorie Grene, 44–62. Chicago: University of Chicago Press, 1995.

Descartes, René. *Discours de la méthode*, ed. Étienne Gilson. Paris: J. Vrin, 1930.

——. *Oeuvres complètes*. 13 vols. Ed. Charles Adam and Paul Tannery. Paris, 1897–1913.

——. *Les passions de l'âme*. Ed. Geneviève Rodis-Lewis. Paris: J. Vrin, 1966.

——. *The Passions of the Soul*. Ed. and trans. Stephen Voss. Indianapolis: Hackett, 1989.

——. *The Philosophical Writings of Descartes*. Ed. and trans. John Cottingham, Robert Stoothoff, and Dugald Murdoch (with Anthony Kenny). 3 vols. Cambridge: Cambridge University Press, 1985–91.

——. *Treatise of Man*. Trans. Thomas Steele Hall. Cambridge: Harvard University Press, 1972.

Duncan, David Allen. "Persuading the Affections: Rhetorical Theory and Mersenne's Advice to Harmonic Orators." In *French Musical Thought, 1600–1800*, ed. Georgia Cowart, 149–75. Ann Arbor: U.M.I. Research Press, 1989.

Elias, Norbert. *The Civilizing Process*. Trans. Edmund Jephcott. Oxford: Blackwell, 1994.

Emerson, J. P. "Nothing Unusual Is Happening." In *Human Nature and Collective Behavior*, ed. T. Shibutani, 208–22. Englewood Cliffs: Prentice-Hall, 1970.

Faret, Nicolas. *L'Honeste homme: ou, l'Art de plaire à la cour*. Lyons: A. Cellier, 1661.

Fox, Christopher. "Introduction: How to Prepare a Noble Savage: The Spectacle of Human Science." In *Inventing Human Science: Eighteenth-Century Domains*, ed. Christopher Fox, Roy Porter, and Robert Wokler, 1–30. Berkeley and Los Angeles: University of California Press, 1995.

Freudenthal, Gad. "Clandestine Stoic Concepts in Mechanical Philosophy: The Problem of Electrical Attraction." In *Renaissance and Revolution: Humanists, Scholars, Craftsmen and Natural Philosophers in Early Modern Europe*, ed. J. V. Field and Frank A. J. L. James, 161–72. Cambridge: Cambridge University Press, 1993.

Fumaroli, Marc. *L'âge de l'éloquence: Rhétorique et "res literaria" de la Renaissance au seuil de l'époque classique*. Geneva: Librairie Droz, 1980.

——. "Rhétorique et philosophie dans le Discours." In *Problématique et réception du Discours de la Méthode et des essais*, ed. Henry Méchoulan, 31–46. Paris: J. Vrin, 1988.

Funkenstein, Amos. *Theology and the Scientific Imagination from the Middle Ages to the Seventeenth Century*. Princeton: Princeton University Press, 1986.

Garber, Daniel. *Descartes' Metaphysical Physics*. Chicago: University of Chicago Press, 1992.

Gaukroger, Stephen. *Descartes: An Intellectual Biography*. Oxford: Clarendon Press, 1995.

Godwin, Joscelyn. *Robert Fludd: Hermetic Philosopher and Surveyor of Two Worlds*. London: Thames and Hudson, 1979.

Goffman, Erving. *The Presentation of Self in Everyday Life*. Garden City, NY: Doubleday, 1959.

Gouhier, Henri. *Les premières pensées de Descartes: Contribution à l'histoire de l'anti-renaissance*. Paris: J. Vrin, 1958.

Greenblatt, Stephen. *Renaissance Self-Fashioning: From More to Shakespeare*. Chicago: University of Chicago Press, 1980.

Hannaway, Owen. "Laboratory Design and the Aim of Science: Andreas Libavius versus Tycho Brahe." *Isis* 77 (1986): 585–610.

Harth, Erica. *Cartesian Women: Versions and Subversions of Rational Discourse in the Old Regime*. Ithaca: Cornell University Press, 1992.

——. *Ideology and Culture in Seventeenth-Century France*. Ithaca: Cornell University Press, 1983.

Hatfield, Gary. "Descartes' Physiology and Its Relation to His Psychology." In Cottingham, 335–70.

Jaynes, Julian. "The Problem of Animate Motion in the Seventeenth Century." *Journal of the History of Ideas* 31 (1970): 219–34.

Keohane, Nannerl O. *Philosophy and the State in France: The Renaissance to the Enlightenment.* Princeton: Princeton University Press, 1980.

Kessler, Eckhard. "The Intellective Soul." In Schmitt et al., 485–534.

Le Clerc, Sébastien. *Pratique de la geometrie, sur le papier et sur le terrain.* Paris: Jombert, 1682.

Levi, Anthony. *French Moralists: The Theory of the Passions 1585 to 1649.* Oxford: Clarendon Press, 1964.

Lindeboom, G. A. *Descartes and Medicine.* Amsterdam: Editions Rodopi NV, 1978.

Maurice, Klaus, and Otto Mayr, eds. *The Clockwork Universe: German Clocks and Automata, 1550–1650.* Washington: Smithsonian Institution; New York: N. Watson Academic Publications, 1980.

Mayr, Otto. *Authority, Liberty, and Automatic Machinery in Early Modern Europe.* Baltimore: Johns Hopkins University Press, 1986.

McNeill, William H. *Keeping Together in Time: Dance and Drill in Human History.* Cambridge: Harvard University Press, 1995.

———. *The Pursuit of Power: Technology, Armed Force, and Society since* A.D. *1000.* Chicago: University of Chicago Press, 1982.

Milton, John R. "Laws of Nature in the Seventeenth Century." In *The Cambridge History of Seventeenth-Century Philosophy,* ed. Michael Ayers and Daniel Garber. Cambridge: Cambridge University Press, forthcoming.

———. "The Origin and Development of the Concept of the 'Laws of Nature.'" *Archives européennes de sociologie* 22 (1981): 173–95.

Montagu, Jennifer. *The Expression of the Passions: The Origin and Influence of Charles Le Brun's* Conférence sur l'expression générale et particulière. New Haven: Yale University Press, 1994.

Moran, Bruce T. "Princes, Machines and the Valuation of Precision in the 16th Century." *Sudhoffs Archiv* 61 (1977): 209–28.

Morgan, Vance G. *Foundations of Cartesian Ethics.* Atlantic Highlands, NJ: Humanities Press, 1994.

Motley, Mark. *Becoming a French Aristocrat.* Princeton: Princeton University Press, 1990.

Muchembled, Robert. *L'invention de l'homme moderne.* Paris: Fayard, 1988.

———. 1991. "The Order of Gestures: A Social History of Sensibilities under the Ancien Régime in France." In Bremmer and Roodenburg, 129–51.

Noreña, Carlos G. *Juan Luis Vives and the Emotions.* Carbondale: Southern Illinois University Press, 1989.

Oakley, Francis. *Omnipotence, Covenant, and Order: An Excursion in the History of Ideas from Abelard to Leibniz.* Ithaca: Cornell University Press, 1984.

Oestreich, Gerhard. *Neostoicism and the Early Modern State.* Cambridge: Cambridge University Press, 1982.

Osler, Margaret. *Divine Will and the Mechanical Philosophy: Gassendi and Descartes on Contingency and Necessity in the Created World.* Cambridge: Cambridge University Press, 1994.

Park, Katharine. "The Organic Soul." In Schmitt et al., 464–84.

Parker, Geoffrey. *The Military Revolution: Military Innovation and the Rise of the West, 1500–1800.* Cambridge: Cambridge University Press, 1988.

Pascal, Blaise. *Oeuvres complètes.* Ed. Louis Lafuma. Paris: Éditions du Seuil, 1963.

Perrault, Charles. *Le labyrinthe de Versailles.* 1677. Reprint, with an afterword by Michel Conan, Paris: Editions du Moniteur.

Porter, Theodore M. *Trust in Numbers: The Pursuit of Objectivity in Science and Public Life.* Princeton: Princeton University Press, 1995.

Price, Derek J. de Solla. "Automata and the Origins of Mechanism and Mechanistic Philosophy." *Technology and Culture* 5 (1964): 9–23.

Revel, Jacques. "La cour." Paper presented at a Center for 17th and 18th Century Studies and William Andrews Clark Memorial Library, UCLA, workshop on "Civility, Court Society and Scientific Discourse: Reframing the Scientific Revolution," 22–23 November 1991.

——. "The Uses of Civility," trans. Arthur Goldhammer. In *A History of Private Life*, ed. Roger Chartier, 3:167–205. Cambridge: Harvard University Press, 1989.

Rodis-Lewis, Geneviève. "Descartes' Life and the Development of His Philosophy." In Cottingham, 21–57.

——. Introduction to Descartes, *Passions.*

——. "Limitations of the Mechanical Model in the Cartesian Conception of the Organism." In *Descartes: Critical and Interpretive Essays*, ed. Michael Hooker, 152–70. Baltimore: Johns Hopkins University Press, 1978.

——. *L'oeuvre de Descartes.* 2 vols. Paris: J. Vrin, 1971.

Roodenburg, Herman. "The 'Hand of Friendship': Shaking Hands and Other Gestures in the Dutch Republic." In Bremmer and Roodenburg, 152–89.

Rorty, Amélie Oksenberg. "Cartesian Passions and the Union of Mind and Body." In *Essays on Descartes' Meditations*, ed. idem, 513–34. Berkeley and Los Angeles: University of California Press, 1986.

Rosenfield, Leonora Cohen. *From Beast-Machine to Man-Machine: Animal Soul in French Letters from Descartes to La Mettrie.* 1941. Reprint, New York: Octagon Books, 1968.

Ross, Stephanie. "Painting the Passions: Charles Le Brun's *Conférence sur l'expression.*" *Journal of the History of Ideas* 45 (1984): 25–47.

Rossi, Paolo. "Hermeticism, Rationality and the Scientific Revolution." In *Reason, Experiment, and Mysticism in the Scientific Revolution*, ed. M. L. Righini Bonelli and William R. Shea, 247–73. New York: Science History Publications, 1975.

Ruby, Jane E. "The Origins of Scientific 'Law.'" *Journal of the History of Ideas* 47 (1986): 341–59.

Sarasohn, Lisa T. "Motion and Morality: Piere Gassendi, Thomas Hobbes and the Mechanical World-View." *Journal of the History of Ideas* 46 (1985): 363–79.

Saunders, Jason Lewis. *Justus Lipsius: The Philosophy of Renaissance Stoicism.* New York: Liberal Arts Press, 1955.

Schama, Simon. *The Embarrassment of Riches: An Interpretation of Dutch Culture in the Golden Age.* Berkeley and Los Angeles: University of California Press, 1988.

Schmitt, Charles B., Quentin Skinner, Eckhard Kessler, and Jill Kraye, eds. *The Cambridge History of Renaissance Philosophy.* Cambridge: Cambridge University Press, 1988.

Sebba, Gregor. "Adrein Baillet and the Genesis of His *Vie de M. Des-Cartes.*" In *Problems of Cartesianism*, ed. Thomas M. Lennon, John M. Nicholas, and John W. Davis, 9–60. Kingston: McGill-Queen's University Press, 1982.

Shapin, Steven. "The House of Experiment in Seventeenth-Century England." *Isis* 79 (1988): 373–404.

———. "'The Mind Is Its Own Place': Science and Solitude in Seventeenth-Century England." *Science in Context* 4 (1991): 191–218.

Shapin, Steven, and Simon Schaffer. *Leviathan and the Air-Pump: Hobbes, Boyle, and the Experimental Life.* Princeton: Princeton University Press, 1985.

Shea, William R. *The Magic of Numbers and Motion: The Scientific Career of René Descartes.* Canton, Mass.: Science History Publications, 1991.

Sloan, Phillip R. "Descartes, the Sceptics, and the Rejection of Vitalism in Seventeenth-Century Physiology." *Studies in History and Philosophy of Science* 8 (1977): 1–28.

Snyders, Georges. *La pédagogie en France aux XVIIe et XVIIIe siècles.* Paris: Presses Universitaires de France, 1965.

Sutton, Geoffrey V. *Science for a Polite Society: Gender, Culture, and the Demonstration of Enlightenment.* Boulder, Colo.: Westview Press, 1995.

Thoren, Victor E. *The Lord of Uraniborg: A Biography of Tycho Brahe.* Cambridge: Cambridge University Press, 1990.

Vartanian, Aram. "Man-Machine from the Greeks to the Computer." In *Dictionary of the History of Ideas*, ed. P. P. Wiener, 3: 131–46. New York: Scribner's, 1973.

Verbeek, Theo. *Descartes and the Dutch: Early Reactions to Cartesian Philosophy 1637–1650.* Carbondale: Southern Illinois University Press, 1992.

———. "Regius's *Fundamenta physices.*" *Journal of the History of Ideas* 55 (1994): 533–51.

Vigarello, Georges. "The Upward Training of the Body." In *Fragments for a History of the Human Body*, ed. Michel Feher et al., 2: 148–96. New York: Zone Books, 1989.

Vives, Juan Luis. *The Passions of the Soul: The Third Book of De anima et viva.* Trans. with an introduction by Carlos G. Noreña. Studies in Renaissance Literature, vol. 4. Lewiston: Edwin Mellen Press, 1990.

———. *Vives: On Education, a Translation of the De tradendis disciplinis.* Trans. Foster Watson. Cambridge: Cambridge University Press, 1913.

Westman, Robert S. "Huygens and the Problem of Cartesianism." In *Studies on Christiaan Huygens: Invited Papers from the Symposium on the Life and Work of Christiaan Huygens, Amsterdam, 22–25 August 1979*, ed. H. J. M. Bos, M. J. S. Rudwick, H. A. M. Snelders, and R. P. W. Visser, 83–103. Lisse: Swets and Zeitlinger, 1980.

Zilsel, Edgar. "The Genesis of the Concept of Physical Law." *Philosophical Review* 51 (1942): 245–79.

Zur Lippe, Rudolph. *Naturbeherrschung am Menschen.* 2d edn 2 vols. Frankfurt am Main: Suhrkamp, 1981.

5

The Revolution in Natural History

Natural History and the Emblematic World View

William B. Ashworth, Jr

Originally appeared as "Natural history and the emblematic world view," in *Reappraisals of the Scientific Revolution*, edited by David C. Lindberg and Robert S. Westman (Cambridge: Cambridge University Press, 1990): 303–32.

The notes in this chapter have been abridged. For complete footnotes the original publication should be referred to.

Editor's Introduction

Traditionally historians of the Scientific Revolution have focused on astronomy and physics. After all, these were the areas in which major changes that warranted the term revolutionary could easily be identified. Other fields somehow failed to measure up and were notable more for their failings and absences: they did not adopt mathematics, they did not have a towering figure such as Newton who transformed the field, and so on. One such field was natural history. In early modern Europe natural history was a very different activity from natural philosophy. The former described and catalogued natural phenomena, the latter sought their causes.

Here William Ashworth notes that the general impression one gains from the historiography is that there was little change in natural history between the middle of the sixteenth century and the end of the seventeenth. But, he argues, the problem with the field is not simply one of neglect. Much of the historian's craft is about fashioning questions. Clearly if historians ask different questions, the artifact they produce will be very different. Ashworth believes that in the historiography of natural history we have been asking the wrong questions. If you are concerned primarily with questions such as who was the first to do something (what Hooykaas terms the progressionist or teleological method, and other historians call the Whig approach to history), you will get a very different kind of answer to the question "what did that person think he was doing and why was he doing it?" That is, we need to understand motives, and since motives only make sense in a particular cultural setting, then we need to examine their cultural setting too.

Sixteenth-century natural historians were embedded in what Ashworth terms the emblematic world view, which flourished during the Renaissance. According to this, the world was a "complex web of associations," of similitudes, sympathies, and antipathies between animals, plants, and minerals. The task of the natural historian was to catalog these associations. But the sources he used were overwhelmingly literary; the natural historian gained his knowledge of natural objects from texts, not from an examination of the objects themselves. By the middle of the seventeenth century, Ashworth argues, the emblematic world view had collapsed.

Certainly the encounter with the New World played a major role in this. Here we have a nice example of how the historical account is refined by later research. Hooykaas suggested that the importance of the discoveries was that they showed the ancients were wrong. Ashworth refines this to show that the importance of New World natural histories was not simply that the ancients were wrong and thus of dubious authority, but that the new animals and plants had no known similitudes.

But there was also a fundamental shift in the self-conception of the natural historian; he came to see his task as filtering out the true from the false, and to do this by an examination of the thing itself. Ashworth suggests this is as much due to natural history's encounter with antiquarianism, with its "obsession" with a truth which can only be reached through the thing itself, as its encounter with the New World. We may see this later self-conception as the only obviously correct one, but Ashworth shows us it is one that must be accounted for historically.

We have here an alternative narrative of the Scientific Revolution, or at least of another revolution, one that was prior to and independent of the better studied one of Baconian empiricism and Cartesian mechanism.

Natural History and the Emblematic World View

William B. Ashworth, Jr

Natural history occupies a shallow niche in most accounts of the Scientific Revolution. One cannot claim that it is totally overlooked, for the typical survey usually contains a chapter on the new herbals of Otto Brunfels and Leonhard Fuchs and the zoological encyclopedia of Conrad Gesner. Pierre Belon's treatise on birds and Guillaume Rondelet's study of fish are usually discussed, and Belon's woodcut comparing the skeleton of a chicken with that of a human is invariably reproduced. But the subsequent period between 1560 and 1660 is either ignored or belittled. Passing attention is sometimes given to Andrea Cesalpino and his attempts at classification; Ulisse Aldrovandi occasionally gets a nod; one of the New World natural histories may be singled out for comment. But this treatment is perfunctory, at best, and many influential figures such as Joannes Jonston are not mentioned at all. Such accounts give the impression that natural history had a brief golden age in the decades between 1530 and 1560 and then stagnated, changing little in the next one hundred years. The implication, then, is that natural history played no formative role in those collective developments that we call the Scientific Revolution. Most historians seem to feel that the natural sciences became important only after 1660, during the era of John Ray, Edward Tyson, and the Paris school of comparative anatomists, and then only because the revolution in the physical sciences had finally begun to be assimilated by natural scientists. As a consequence, the intervening one-hundred-year period between Gesner and Ray is almost totally neglected, a neglect that extends far beyond the survey level of scholarship. Implicit in this neglect are the assumptions that natural history did not change between 1560 and 1660, that Aldrovandi, Gesner, and Jonston were all engaged in much the same kind of activity, and that it is an activity not really worth further study.[1]

Distorted Perceptions of the Role of Natural History in the Scientific Revolution

I believe that our assumptions and conclusions concerning the nature of natural history are seriously flawed and have prevented us from understand-

1 Since it is hard to know whom to blame for our distorted view of natural history's role in the Scientific Revolution, I prefer not to blame anyone, or rather, to blame us all. Survey writers, after all, must rely on secondary scholarship, and in the field of natural history they have not had

ing a crucial development in late Renaissance and early seventeenth-century thought. Before I attempt to demonstrate a more fruitful approach, I would like to suggest a reason why such assumptions have persisted, even though they may, in fact, be dead wrong. The problem seems to be that we have not, in recent years, reexamined our presuppositions about how one should write the history of the natural sciences, at least for the Renaissance period. This is surprising, since in the past three decades we have thoroughly reworked the historiographic principles we employ when writing about Copernicanism, or the mechanical philosophy, or practically any development in the physical sciences. Such retooling has not occurred for late Renaissance natural history; we still follow the lead of earlier historians, such as Charles Singer, F. J. Cole, and Erik Nordenskiold, who were primarily interested in such questions as who discovered the fish bladder or who first classified the bat with the mammals – historians, in short, who were looking for the origins of biology, and, if not that, then at least for the roots of modern zoology and botany.[2] Now, if one is looking for new discoveries about the chameleon, then it is natural to jump from Belon, who first drew it correctly, to the Paris school, which first took it apart to reveal its anatomical structure. If one has an interest in classification, it is natural to mention Cesalpino and then proceed directly to Ray, who was the first to make much of an advance beyond Aristotle.

I suspect that modern survey writers will passionately deny, with some justification, that they write history as F. J. Cole did, but when one reads the recent literature, the same assumptions are implicitly present. Gesner is lauded for his attempt to gather firsthand information and for his illustrations; he is chided for his humanist fondness for philology and for his lack of any critical sense. Aldrovandi is lightly praised for his anatomical investigations and then dismissed for his unchecked tendency to include biologically irrelevant material, such as fables and proverbs. Jonston is ignored because he seems to be only a truncated Aldrovandi. Rondelet and Belon, on the other hand, and perhaps Volcher Coiter, are given space far exceeding their contemporary importance, because they studied specimens firsthand, dissected them, drew them from life, and scorned the humanistic apparatus of Gesner; in short, because they practiced something faintly resembling biology.

I would like to suggest that our view of natural history has been distorted because we have not been asking ourselves the right questions. The questions we should be asking are these: Why did Renaissance scholars gather and publish information about the natural world? What kind of material was

much to read in recent years. There are exceptions, which I will gratefully acknowledge as we proceed.
2   I refer here to Charles Singer, *A Short History of Biology* (New York: Schuman, 1931, subsequently revised); F. J. Cole, *A History of Comparative Anatomy from Aristotle to the Eighteenth Century* (London: Macmillan, 1944); and Erik Nordenskiold, *The History of Biology: A Survey*, trans. Leonard B. Eyre (New York: Tudor, 1946).

included in their compilations, and why? What was the intended audience of the publications; what was the intended use of the information contained in them? Was the study of nature part of some larger cultural endeavor; did it receive encouragement from patrons and princes; and if so, why? If Gesner and Aldrovandi were trying to write biological textbooks and failed in the attempt, then they may merit the criticism they have received. But if they were trying to do something quite different, perhaps we should first try to understand their motives, and the cultural setting of such motives, before dismissing their efforts so readily.

What I would like to do in this essay is to ask some of these questions, at least in preliminary form, and see where they lead us. I will limit my inquiry to the zoological side of natural history and will focus on the period 1550 to 1650. I hope to show, even in this brief reappraisal, that when we look at natural history through contemporary eyes, we see an entirely different world from ours, a world where animals are just one aspect of an intricate language of metaphor, symbols, and emblems. This "emblematic world view," as I choose to call it, was the single most important factor in determining the content and scope of Renaissance natural history.[3] Moreover, the nature of this world of symbols and correspondences changed considerably between 1550 and 1650; it grew considerably rich between Gesner and Aldrovandi, and dissipated completely by the time of Jonston. Viewed from this perspective, the natural histories of Gesner, Aldrovandi, and Jonston were markedly different, not stamped from the same mold. Most important, I hope to show that the demise of emblematic natural history was a crucial part of the development that we call the Scientific Revolution. It was not simply an aftermath of Descartes and the mechanical philosophy but an independent, and perhaps even broader, cultural shift that had profound consequences for the evolution of seventeenth-century science.

Gesner and Humanist Natural History

Perhaps the best way to open a window onto the emblematic world of the Renaissance is to open the *History of Animals* of Conrad Gesner (1516–1565) and read an article, with no expectations or preconceptions – to let the world reveal itself. What might one learn, for example, by consulting his chapter on *pavo*, the peacock? The article begins with an attractive woodcut, followed by a list of the bird's names in different languages, and a description, pieced together from ancient authorities such as Aristotle and Pliny.

3   My choice of the term "emblematic world view" is discussed, and defended, in note 26, this chapter.

Attention is then given to the peacock's habits and characteristics, where we learn, for example, that its flesh does not decay after death and that it is ashamed of its feet. On subsequent pages we encounter a discussion of all known peacock adjectives and their origins, such as "peacock blue," or the Peacock River in India, or the "peacock stone." We are told that the peacock was associated with the goddess Juno and appeared with her on ancient coins, and we are treated to several fables involving the pair. We are informed of the myth of Argus, who had one hundred eyes, which were transformed, after his death, into the spots on the peacock's tail. We also encounter peacock proverbs, peacock recipes, peacock medicines, and peacock legends. Every single statement is supported by a named authority, usually classical, but often contemporary. Gesner has provided us with the ultimate peacock concordance.[4]

Now, if what you seek is a collection of true statements about the peacock, or an anatomical description, or the peacock's place in a taxonomic scheme based on physical characteristics, then you are bound to be disappointed by Gesner's account. But if you are interested in confronting, in one place, that complex web of associations that links the peacock with history, mythology, etymology, the rest of the animal kingdom, indeed with the entire cosmos, then you are certain to be richly rewarded. Gesner believed that to know the peacock, you must know its associations – its affinities, similitudes, and sympathies with the rest of the created order.[5] Michel Foucault has suggested that this search for similitudes and resemblances was the principal guiding episteme for all of Renaissance thought, and, in the case of Gesner's natural history, he was absolutely right.[6]

From what sources does Gesner assemble his peacock network of associations? Some of them are well known to students of natural history, and we will not linger over these: Aristotle, of course, and Pliny, along with Aelian, Plutarch, Theophrastus, Varro, and practically every other classical writer who discussed animals. We would expect to find them in a compilation written by a humanist as knowledgeable as Gesner. But what are the names of Erasmus, Du Choul, and Horapollo doing in the margins? Ovid, Alciati, and the Greek *Florilegia*? What works do their names represent, and what do they have to do with natural history? Since these sources form an important part of Gesner's world view that is little acknowledged by historians of science, an accounting seems in order.

---

4 Conrad Gesner, *Historica animalium Lib III: De avium* (Zurich: Froschover, 1555), pp. 630–9.
5 There is a great deal of literature on Gesner, but most of it fails to deal with the questions raised; [. . .]
6 Michel Foucault, *The Order of Things: An Archaeology of the Human Sciences* (New York: Pantheon Books, 1970), pp. 17–45. [. . .]

The Cultural Matrix of Sixteenth-century Natural History

There seem to have been six developments in sixteenth-century thought that, added to the classical literature on natural history, determined the cultural matrix of late Renaissance natural history. We might call these, for convenience, the *hieroglyphic, antiquarian, Aesopic, mythological, adagial,* and *emblematic* traditions. The number six is not intended to be canonical; we could just as easily organize them into five or ten groups, since all of these traditions were densely interwoven, but the six-part division works well for purposes of discussion.

Hieroglyphics

Renaissance fascination with hieroglyphics began in the early fifteenth century, when the *Hieroglyphics* of Horapollo (dates unknown) was recovered and translated from the Greek. Horapollo's treatise is essentially a dictionary of symbols, of which a large proportion is animal. It reveals, for example, that when the Egyptians drew a pig, it was meant to symbolize a pernicious person, whereas a weasel represented weakness, a fly impudence, and so forth. The humanist mind was fascinated with such revelations, because hieroglyphics seemed to be a language of symbolic images – a language in which understanding is conveyed immediately, much as God understands things, without the mediation of conventional language. Marsilio Ficino, in particular, was vastly impressed with the possibilities of such a Platonic language, and so were many of his followers. Horapollo was first printed in 1505, and the *Hieroglyphics* went through many more editions by the end of the sixteenth century.[7]

The early impact of Horapollo on natural history is best seen in the example of Albrecht Dürer (1471–1528). In 1512 his friend Willibald Pirckheimer translated Horapollo's treatise. Dürer illustrated the manuscript, and although the original is lost, a copy survives, containing Dürer's meticulous depictions of such hieroglyphs as a dog wearing a stole (representing the judgment of kings) and a lion (representing fear). More interesting, shortly thereafter Dürer designed a large triumphal arch for Maximilian I, at the top of which sits the emperor, surrounded by symbolic animals: the lion, the dog with stole, a crane on raised foot (a guard against enemies), a bull (courage with temperance), and others. Pirckheimer himself then "translated" these Horapollonian images into a message in praise of the emperor.[8]

7 The best introduction to Horapollo is George Boas's introduction to his edition, *Hieroglyphics of Horapollo,* Bollingen series, no. 23 (New York: Pantheon Books, 1950). [. . .]
8 Erwin Panofsky, *The Life and Art of Albrecht Dürer* (Princeton: Princeton University Press, 1943), pp. 173–77. The Dürer drawings and woodcuts after Horapollo are reproduced in Karl

The effect of the hieroglyphic revival on natural history was immediate and profound. Weasels, cranes, and lions became part of a visual language; they were symbols, but even more, they were Platonic ideas, whose meaning the mind could immediately perceive. Animals were living characters in the language of the Creator, and the naturalist who did not appreciate or understand this had failed to comprehend the pattern of the natural world.

Antique coins and Renaissance medals

Closely related to the interest in hieroglyphics was the Renaissance fascination with antiquities, especially medals and coins. Antique Roman coins typically had a portrait on one side, and, on the reverse, an image that seemingly had symbolic meaning. The coins of Titus Vespasian, for example, showed a dolphin twined around an anchor. Renaissance humanists, already by the mid-fifteenth century, began to devise medals in imitation of ancient coins, and here again the impulse seems to have come from a fascination with symbolism. Leon Battista Alberti, in 1438, graced his medallic reverse with an eye surrounded by a laurel wreath, with the motto, from Cicero, "Quid tum" (What then?). Pisanello designed a medal for Belloto Cumano, in 1447, that has a weasel or ermine on the reverse, representing purity.[9]

The early medals developed apart from the hieroglyphic tradition, but by the early sixteenth century the two were closely intertwined. Erasmus, in his *Adagia*, which I will discuss shortly, commingled the two; after mentioning that his friend, the printer Aldus Manutius, had taken Vespasian's dolphin and anchor as his own personal device, he says that the symbol means "festina lente" (Make haste slowly), "as the books on hieroglyphics tell us," and he then proceeds to explain the importance of a symbolic language.[10]

Interest in numismatics continued to increase through the middle of the sixteenth century, when there began to appear the first antiquarian treatises on ancient coinage, filled with plate after plate of symbols and mottoes, many of them animal. The most important of these compilations were Aeneas Vico, *Images of Emperors from Antique Coins* (1553); and Guillaume Du Choul, *Religion of the Ancient Romans* (1556). Works such as these were very important sources for late Renaissance humanists, because they were based on artifacts,

Giehlow, "Die Hieroglyphenkunde des Humanismus in der Allegorie der Renaissance," *Jahrbuch der kunsthistorischen Sammlungen der Allerhöchsten Kaiserhauses*, 32 (1915):1–232; frontispiece and pp. 170–218.
9   Renaissance fascination with coin symbolism is discussed in Don C. Allen, *Mysteriously Meant: The Rediscovery of Pagan Symbolism and Allegorical Interpretation in the Renaissance* (Baltimore: Johns Hopkins University Press, 1970), esp. pp. 249–76. The Alberti and Cumano medals are discussed in Wittkower, "Hieroglyphics," pp. 120–2. [. . .]
10   See the reference in note 16, this chapter.

not written history, and antiquarianism was just then developing as an alternative method of studying the past.[11] Moreover, the frequent appearance of peacocks, lions, and eagles on ancient coins was convincing evidence that the study of antiquities was an important aspect of natural history.

Aesopic fables

The third tradition was that of the fable, especially the Aesopic fable. The collection of fables ascribed to Aesop has a convoluted history; it came down to the Renaissance in verse and prose forms, in Greek and Latin versions, and with varying numbers of fables. For our purposes, it suffices to note that one version, printed around 1476, was rapidly translated into vernacular languages and was reprinted constantly throughout the Renaissance.[12] Particularly nice editions were published in Paris in 1547, with illustrations by Bernard Salomon, and in 1567, with illustrations by Marcus Gheeraerts; both were often reissued.[13] So when sixteenth-century humanist naturalists became interested in the symbolic meanings of animals, the Aesopic corpus became an important source. No student of the peacock would want to ignore the fable of Juno and the peacock, in which the peacock complains that he does not have a voice like the nightingale, because there is a moral here for those who are not content with their station in life.

Classical mythology

The fourth tradition that made an important contribution to the multilayered world view of the Renaissance was the mythological. It is well known that classical mythology had an overwhelming impact on Renaissance art and literature, but it also left its mark on natural history. Animals, after all, romped around Mount Olympus along with the deities, and it is difficult even today to picture Hera without her owl, Jupiter without his eagle, and Juno without the aforementioned peacock; in the Renaissance it was impossible. The principal

11   Arnaldo Momigliano, "Ancient History and the Antiquarian," in his *Studies in Historiography* (New York: Harper & Row [Harper Torchbooks], 1966), pp. 1–39. This article first appeared in *Journal of the Warburg and Courtauld Institutes*, 13 (1950):285–315.

12   The Aesopic tradition in the Renaissance can be partially unraveled by consulting Ben Edwin Perry's introduction to his translation of *Babrius and Phaedrus* (Loeb Classical Library, 1965), pp. xi–cii, and Joseph Jacobs, *The Fables of Aesop . . . :* vol. 1, *History of the Aesopic Fable* (1889; reprint edition, New York: Franklin, 1970).

13   The title notwithstanding, there is much useful information on sixteenth-century Continental editions of Aesop in Edward Hodnett, *Aesop in England: The Transmission of Motifs in Seventeenth-century Illustrations of Aesop's Fables* (Charlottesville: University Press of Virginia, 1979), pp. 34–50.

source for the zoology of myth was Ovid's *Metamorphoses*, and Gesner was as familiar with this work as he was with Aristotle. In the sixteenth century, however, Ovid was supplemented by other scholarly treatises, most notably those of Lilio Giraldi, Natale Conti, and Vincenzo Cartari, all published around midcentury. The work of Cartari, *Images of the Gods*, published in 1556, was particularly influential as a sourcebook of mythological animal imagery.[14]

Near the end of the sixteenth century, the mythological tradition spawned an offshoot that is best called "iconology," after the master treatise in that genre, Cesare Ripa's *Iconologia*, first published in 1593 and often reissued. Ovid provided attributes for the gods; Ripa provided attributes for personifications of all kind: nature, intellect, envy, modesty, heresy. Many of Ripa's attributes were drawn from natural history; thus a veiled woman with an elephant by her side represents religion; a long-eared woman pointing with her finger and holding a peacock represents arrogance. In the late sixteenth century these animal attributions joined those drawn from Horapollo arid Pliny to create an impressively rich language of associations for the natural world.[15]

Adages and epigrams

There are two traditions left to elucidate, however, and it might be argued that these two are the most important and influential of all. First, and the fifth in our catalog, is the tradition of the adage, or proverb. In the sixteenth century the adage was synonymous with the name of Desiderius Erasmus (1466?–1536). In 1500 Erasmus published a collection of proverbs, the *Adages*, culled from ancient writings and illuminated by his own very personal commentary.[16] The work was enlarged and reprinted continually for almost

14 Lilio Giraldi, *De deis gentium varia et multiplex historia* (Basel, 1548): Natale Conti, *Mythologiae* (Venice, 1551); Vicenzo Cartari, *Le imagini colla sposizione degli dei degli antichi* (Venice, 1556). The best introduction by far to the sixteenth-century mythological tradition is still Jean Seznec, *The Survival of the Pagan Gods: The Mythological Tradition and Its Place in Renaissance Humanism and Art*, trans. Barbara F. Sessions (1953; reprint edition, New York: Harper & Row [Harper Torchbooks], 1961), esp. pp. 219–56. See also Don C. Allen, *Mysteriously Meant*, pp. 201–47.
15 Cesare Ripa, *Iconologia* (Rome, 1593; first illustrated edn, Rome, 1603). On the importance of Ripa, both for the Renaissance and for the modern scholar, see D. J. Gordon, "Ripa's Fate," in '*The Renaissance Imagination*, ed. Stephen Orgel (Berkeley and Los Angeles: University of California Press, 1975), pp. 51–74.
16 Desiderius Erasmus, *Adagiorum collectanea* (Paris: Phillip, 1500); *Adagiorum chiliades* (Venice: Aldi, 1508); *Adagiorum chiliades* (Basel: Froben, 1536). There were many other editions; see Margaret Mann Phillips, *The "Adages" of Erasmus: A Study with Translations* (Cambridge: Cambridge University Press, 1964). An English translation of the entire collection is now in publication: Desiderius Erasmus, *Adages*, trans. Margaret Mann Phillips, annot. R. A. B. Mynors (Toronto: University of Toronto Press, 1982–), but so far only two volumes – containing the first one thousand adages – of the projected seven have appeared.

four decades, and by the last edition there were over forty-one hundred adages in the collection, the total collected aphoristic wisdom of antiquity. Many of them are still quite familiar: "Omnem movere lapidem" (Leave no stone unturned) or "Ligonem ligonem vocat" (Call a spade a spade).[17] And many of them concern animals. Thus Erasmus tells us: "Multa novit vulpes, Echinus vero unum magnum" (The fox knows many ways [to survive]; the hedgehog one great one), referring to the hedgehog's sole but effective defense of rolling up into a ball. Erasmus's compendium of proverbs was one of the most widely influential works of the entire sixteenth century; Gesner, in particular, seems to have read the entire work most carefully.[18]

One body of ancient writings that Erasmus drew on should be singled out, since some would argue that it has a separate life of its own: the so-called Greek Anthology. This collection of ancient epigrams was assembled by Planudes in the thirteenth century; it was first printed in Greek in 1494, and in a number of Latin editions after 1520. The Greek Anthology contains few epigrams that concern animals – Aesop had more or less cornered this market – but it helped create a taste for the clever, pithy aphorism that, by the middle of the sixteenth century, spread to include observations about the natural world.[19]

Emblems and devices

The sixth and last tradition that I wish to single out is the emblematic. The emblem was one of the most influential creations of the late Renaissance.[20] The original intention of the inventor, Andrea Alciati (1492–1550), was to devise epigrams that were especially enigmatic, so that readers would get a sudden and pleasing illumination when they figured them out, with the help

17 Margaret Mann Phillips, *Erasmus on His Times: A Shortened Version of the "Adages" of Erasmus* (Cambridge: Cambridge University Press, 1967), pp. xi–xii.
18 Rosalie Colie, in her wonderful lecture "Small Forms: Multo in parvo," explains the popularity of adages, calling them "keys to culture, or convenient agents of cultural transfer" and suggesting that an adage "compresses much experience into a very small space; and by that very smallness makes its wisdom so communicable"; see Rosalie L. Colie, *The Resources of Kind: Genre-theory in the Renaissance*, ed. Barbara K. Lewalski (Berkeley and Los Angeles: University of California Press, 1973), pp. 32–75, esp. pp. 33–4.
19 For a thorough study of the influence of the Greek Anthology in the sixteenth century, see two books by James Hutton: *The Greek Anthology in Italy to the Year 1800* (Ithaca, NY: Cornell University Press, 1935) and *The Greek Anthology in France and in the Latin Writers of the Netherlands to the Year 1800* (Ithaca, NY: Cornell University Press, 1946).
20 The literature of Renaissance emblematics is vast, although much of it is highly specialized. The best introduction is still the first volume of Mario Praz, *Studies in Seventeenth-century Imagery*, 2 vols, Studies of the Warburg Institute, no. 3 (London: Warburg Institute, 1939–1947); vol. 2 is a bibliography. [. . .]

of a commentary; an accompanying image was not intended. But when Alciati's *Emblemata* was first published in 1531, woodcut illustrations were added, and by midcentury the visual image had become an indispensable part of the emblem. The emblem proper ultimately came to consist of three parts: a visual image, a short motto, and a slightly longer epigram. In the ideal emblem, each element was necessary, but not sufficient, for comprehension; taken together, they provided a pleasing and useful insight.[21] A pleasant example, taken from a late emblem book, shows a peacock gazing at its feet, in defiance of Pliny's claim, with the wonderful motto: "Nosce te ipsum" (Know thyself).[22]

Closely related to the emblem was the "device," or *impresa*. A device was a sort of personal emblem, with an image and motto particularly appropriate to the owner, and the device actually predates the emblem by half a century, originating as a badge worn in battle. But emblems and devices rode to ascendancy in tandem in the sixteenth century, and in the late sixteenth century personal devices were often expanded into emblems, and emblems were converted into devices. A good example of an animal device that acquired general circulation is that of King Louis XII of France, which showed a porcupine with the motto "Cominus et eminus" (Hand to hand and from afar), cleverly suggesting that the king, like a porcupine, can triumph in battle as well as as by diplomatic action.[23]

The emblem tradition blossomed in a manner that is almost unimaginable to the modern student who is unfamiliar with it. Alciati's book went through dozens of expansions and reissues, and these spawned, in turn, a proliferating host of rivals. By 1600 there were hundreds of different emblem treatises in print, and production continued unabated for several more decades, finally beginning to slacken off only after 1650. One reason why the emblem tradition was so important was that it brought together most of the other traditions we have outlined. The emblem is clearly an outgrowth of the love for proverbial wisdom, and Alciati was very much influenced by Erasmus and the Greek Anthology.[24] Hieroglyphics played an important role in the development of the emblem, as did the mottoes and images on ancient coins and the moral lessons from Aesop and Ovid.[25] Because of the unifying character of the emblematic

21 A variorum translation of Alciati has recently been published: Peter M. Daly, ed.; with Virginia W. Callahan, assisted by Simon Cuttler, *Andreas Alciatus*: vol. 1, *The Latin Emblems Indexes and Lists*; vol. 2, *Emblems in Translation* (Toronto: University of Toronto Press, 1985). [. . .]
22 Peter Ieslburg, *Emblemata politica* (Nuremberg, 1617), no. 3.
23 The device can be found in Paolo Giovio, *Dialogo dell'Imprese militari et amorose* (Lyons, 1559), p. 20, and in many subsequent device and emblem books.
24 Hessel Miedema, "The Term *emblema* in Alciati," *Journal of the Warburg and Courtauld Institutes*, 31 (1968):234–50; Alison Saunders, "Alciati and the Greek Anthology," *Journal of Medieval and Renaissance Studies*, 12 (1982):1–18.
25 Daniel Russell, "Emblems and Hieroglyphics: Some Observations on the Beginnings and Nature of Emblematic Forms," *Emblematica*, 1 (1986):227–43.

tradition, and because of the fact that it struck such a resounding chord in late Renaissance thought, I have called the mental outlook that welcomed it the "emblematic world view."[26]

The Emblematic World View

The emblematic world view is, in my opinion, the single most important factor in determining late Renaissance attitudes toward the natural world, and the contents of their treatises about it. The essence of this view is the belief that every kind of thing in the cosmos has myriad hidden meanings and that knowledge consists of an attempt to comprehend as many of these as possible. To know the peacock, as Gesner wanted to know it, one must know not only what the peacock looks like but what its name means, in every language; what kind of proverbial associations it has; what it symbolizes to both pagans and Christians; what other animals it has sympathies or affinities with; and any other possible connection it might have with stars, plants, minerals, numbers, coins, or whatever. Gesner included all this, not because he was uncritical or obtuse, but because knowledge of the peacock was incomplete without it. The notion that a peacock should be studied in isolation from the rest of the universe, and that inquiry should be limited to anatomy, physiology, and physical description, was a notion completely foreign to Renaissance thought.

Once the modern student becomes comfortable with this complex world of symbols and associations and starts to read Renaissance natural histories with more awareness of the way these different discourses interacted, certain developments appear, in this new light, more understandable. We begin to see, for example, why Pierre Belon (1517–1564) and Guillaume Rondelet (1507–1566) did not have more impact in the late Renaissance, if they were in fact better zoologists than Gesner. Historians have fumbled for explanations, but it now seems evident that Belon and Rondelet attempted to place animals in a context that was much too limited. Anatomy, physiology, and classification may be the heart of modern zoology, but in the sixteenth century they were only several strands of a much more complex web, and contemporaries obviously felt that such a stripped-down world was incomplete; the zoological

---

26   The choice of the term "emblematic" seems defensible enough, although one could make a good case for "symbolic" (but *not* "magical" or "hermetic," which reflect a serious misunderstanding of the source of these traditions). But why "world view"? Why not "episteme," or "paradigm," or "discourse"? The answer is simply that all of these terms are laden with connotations that say more about twentieth-century historiography than sixteenth-century epistemology." "World view," at least for the moment, seems to mean exactly, and only, what it says. If it buzzes too badly, I would be happy to abandon it for a more acceptable label for the Renaissance outlook I have tried to identify.

world depicted by Belon and Rondelet was not the zoological world inhabited by Renaissance man; it had lost too much of its richness and meaning. Gesner's world, on the other hand, was complex and interwoven, and the success enjoyed by his works and that of his successors is evidence that readers shared and cared for this world of resemblances.

We can also realize what a mistake it is to call the outlook of Gesner and his followers "medieval," as historians have often done.[27] The adjective crops up because medieval bestiaries also incorporated animal symbolism and morals. But we can now understand that Gesner's symbolism is of quite a different kind and a higher order. Gesner's ancient sources were mainly classical, rather than Christian, and in addition he drew on many contemporary traditions that were unknown to the Middle Ages. It is noteworthy that bestiary symbolism was drawn primarily from the *Physiologus*, and Gesner hardly used the *Physiologus* at all (perhaps because it was not printed until 1587). There are many tales included by Gesner that are also in the *Physiologus*, but that is because both have a common source in Pliny. And Gesner rarely includes the medieval Christian morals that were the core of the bestiary tradition. So Gesner's world view may have been rich in animal symbolism, but there was nothing distinctively "medieval" about it.

Another thing we notice is that the world of associations inhabited by Gesner was something quite different from what is sometimes called the "magical world view." Gesner was indeed familiar with magical treatises, most notably the *Kiranides*, and his discussions of sympathies usually come from such sources, if they were not drawn from Pliny.[28] But they form only a small fraction of his sources and his world view. Magic, or hermetism, has come in for a lot of attention in the last decades and has been offered up by some as *the* world view of the Renaissance, the outlook that was to be replaced by the mechanical philosophy. I merely wish to point out here that in fact magic, or hermetism, was only one element of a much larger picture; only one tradition among dozens that fused to form the emblematic world view.

Aldrovandi and Emblematic Natural History

And finally, we are ready to appreciate the difference between Aldrovandi and Gesner. Ulisse Aldrovandi (1522–1605) must be the most underappreciated

27   Raven, *English Naturalists*, p. 47; Paul Delaunay, *La zoologie au seizième siècle*, Histoire de la pensée, no. 7 (Paris: Hermann, 1962), pp. 63–81; Lynn Thorndike, *A History of Magic and Experimental Science*, 8 vols. (New York: Columbia University Press, 1923–1958), 6:277 (discussing Aldrovandi).
28   Very little attention has been given to the impact of the *Kiranides* – the purported writings of Kiranus, king of Persia – on Renaissance thought, perhaps because the work never saw its way into print. [. . .]

naturalist of the early modern era. His thirteen massive folios stand high and dry on library shelves, like so many beached whales, forbidding in their bulk, alien in their contents, and apparently seldom read. The encyclopedic format has led most historians to conclude that he was just another Gesner, except that he did not know when to stop.[29] This opinion is unfortunate, because Aldrovandi was not "Gesner redivivus." If one concentrates on the biological parts of his compendiums, there are indeed great similarities. But if one reads on for the associations, one discovers that there has been a great change in fifty years. Suppose we turn to Aldrovandi's article on the peacock.[30] We notice, first of all, that it is thirty-one pages long, compared to Gesner's eight. Gesner divided his article into eight sections; Aldrovandi has thirty-three topics in all, and it is well worth listing the titles of these:

aequivoca	aetas	moralia
synonyma	volatus	hieroglyphica
genus	mores	symbola
differentiae	ingenium	proverbia
descriptio	sympathia	usus in sacris icones
locus	antipathia	usus in externis
coitus	corporis affectus	usus in medicina
partus	cognominata	usus in cibis
incubatus	denominata	apologi
educatio	praesagia	fabulosa
vox	mystica	historica

It is one thing to talk about a "web of associations"; it is much more impressive to see this web laid out, strand by strand, as Aldrovandi does. Aldrovandi's network is similar in kind to Gesner's but many times more intricate. What has happened to the emblematic world in the intervening fifty years to swell it to such splendor?

Gesner had compiled his encyclopedias in the 1550s. At that time the adages of Erasmus were in wide circulation, as was the mythology of Ovid, and Gesner utilized both freely. But many of the other traditions were just beginning to flower. Horapollo had been available in print for quite some time, but only with the publication of the *Hieroglyphics* (Basel, 1556) of Piero Valeriano (1477–1558) did fascination with hieroglyphics really begin to spread. So we find in Gesner only passing attention given to hieroglyphic meanings. The great numismatic encyclopedias did not appear until the mid-1550s.

29   Aldrovandi is ignored, or deplored, in virtually every English-language discussion of Renaissance natural history, whether survey or specialized. Fortunately Italian scholars have launched a rescue effort for their beleaguered countryman. [. . .]
30   Ulisse Aldrovandi, *Ornithologia* II (Bologna, 1600), pp, 1–31.

Ripa's *Iconologia* was unavailable to Gesner, as was the printed *Physiologus*. The fable tradition was just catching hold, and most of the best editions of Aesop did not appear until the 1570s. And most important, the emblematic tradition was barely a bud when the first volume of the *History of Animals* lumbered off the presses. Few animal emblems were in circulation in Gesner's day, and although he utilized the ones available, they do not dominate his descriptive associations.

It is the efflorescence of the emblem tradition that marks the biggest difference between Gesner and Aldrovandi, and I would like to demonstrate the growth of animal emblematics before returning to Aldrovandi. Animals did not play a central role in Alciati's *Emblemata*; they were present, but not omnipresent. But when others began composing emblem books, they turned to Horapollo and Piero Valeriano for inspiration, and there animal symbols are abundant. So from 1560 on we begin to see more and more attention given to the epigrammatic meanings of the natural world. This trend culminated in the publication of the *Collection of Symbols and Emblems* of Joachim Camerarius (1534–1598) from 1593 to 1604. This set of four volumes contains four hundred emblems, and every one involves an animal or plant. In the second volume, on quadrupeds, we find emblems for one hundred animals – not only horses and lions, but hedgehogs, ichneumons, chameleons, weasels, and even the New World *simivulpa*, or opossum.

It is important to understand that Camerarius was as much a student of nature as Gesner, and his emblem book was intended as a contribution to natural history, as well as to emblematics. The commentary to his peacock emblem, for example, refers to Aristotle, Pliny, Ovid, Isidore, as well as earlier emblem books, and Camerarius apparently saw no contradiction between his emblem-book production and his botanical work; both illuminated the emblematic world of nature.[31]

By the beginning of the seventeenth century, there was available a cornucopia of animal allegories and symbolism for anyone interested in adding to the traditional animal similitudes. Aldrovandi was very much interested. Just after Camerarius's first volumes of emblems rolled off the press, Aldrovandi began to issue the first volumes of his natural history. The *Ornithology* was the first to appear, in three volumes published between 1599 and 1603. The volume on insects followed. Aldrovandi died, and the production slowed

---

31   Joachim Camerarius, *Symbolorum & emblematum ex re herbaria desumtorum centuria una collecta* (Nuremberg, 1590 [i.e., 1593]); *Symbolorum & emblematum ex animalibus quadrupedibus desumtorum centuria altera collecta* (Nuremberg, 1595); *Symbolorum & emblematum ex volatilibus ex insectis desumtorum centuria tertia collecta* (Nuremberg, 1596); *Symbolorum et emblematum ex aquatilibus et reptilibus desumtorum centuria quarta* (Nuremberg, 1604). The importance of Camerarius for natural history is stressed by Wolfgang Harms," On Natural History and Emblematics in the Sixteenth Century," in *The Natural Sciences and the Arts*, ed. Allan Ellenius, Acta Universitatis Upsaliensis, Figura Nova, no. 22 (Uppsala: Almqvist & Wiksell, 1985), pp. 67–83. [. . .]

slightly, but not much, as his assistants and heirs took over responsibility for bringing the Aldrovandi corpus to light. The first volume on quadrupeds came out in 1616, and subsequent huge volumes plopped into view with intermittent regularity, right up until 1648. Why did Aldrovandi need three volumes on birds and three for quadrupeds, where Gesner had one for each? The reason is that Gesner's humanist text had been swollen by incorporating all of the new contributions of the sixteenth-century students of hieroglyphics, emblems, adages, and antiquities. Let us consider another specimen animal, this time from the world of quadrupeds.

The *echinus*, or hedgehog, was well known to classical authorities; should one look up the entry in Gesner, one would find most of the interesting hedgehog stories gathered together.[32] One would learn that the hedgehog carries home grapes and apples on its spines – never in its mouth. When *echinus* walks, it squeaks like a cartwheel; when a male and a female copulate, they do so face to face. Gesner transcribes two proverbs from Erasmus's *Adages:* One we have already discussed; the other, *Echinus partum differt* (The hedgehog delays childbirth), likens a poor man, who puts off payment of debts, to the hedgehog mother, who tries to retard the delivery of her spiny whelps. Gesner also adds a long section on medicinal uses. But there are no examples drawn from hieroglyphics, emblems, or numismatics.

Aldrovandi's discussion is much more extensive.[33] There is now a lengthy paragraph on *antipathia*, or antipathies, informing us that the hedgehog is a bitter enemy of the wolf, detests serpents, and is not fond of those plants that have spines themselves. Under the heading "Emblemata & Symbola" one can find both of Camerarius's emblems quoted in full, with motto and epigram. Aldrovandi includes a section on *simulacra*, or images, where he reveals his familiarity with Ripa's *Iconology*, pointing out that Ripa's figure of *laesiones*, or oratory, has a hedgehog in one hand.

And in other sections on hieroglyphics, morals, omens, symbols, and so forth, one finds every reference to the hedgehog that is made by Piero Valeriano, Horapollo, the *Physiologus*, Erasmus, and most of the important emblem writers. With all these resources Aldrovandi is able to spin a net of associations and similitudes that is far more complex than anything that Gesner was able to achieve. Aldrovandi's world needed thirteen volumes to contain it.

The emblematic view of nature continued to prevail through the first half of the seventeenth century, periodically refreshed by the appearance of additional Aldrovandi zoological volumes. And while Aldrovandi was a major force in its persistence, other zoologists fashioned similar world views, often independently. The *Historie of Four-Footed Beastes* (1607) by Edward Topsell (1572–1638) provides a good example. Topsell has been much maligned as an unimaginative plagiarist of Gesner, and some of the criticism

---

32   Gesner, *Historia animalium Lib. I. de quadripedibus viviparis* (Zurich, 1551), 1:399–409.
33   Ulisse Aldrovandi, *De quadrupedibus digitatis* (Bologna, 1637), pp, 459–70.

is deserved.[34] But it is of interest that Topsell frequently added new material to that he took from Gesner, and most of it consisted of references drawn from emblematic and hieroglyphic literature. Since Topsell wrote before the appearance of Aldrovandi's volumes on quadrupeds, he must have gleaned this new material on his own, by perusing the works of Camerarius and Piero Valeriano. Moreover, whatever his failings as a zoologist, Topsell knew exactly what he was trying to do in his book. His "Epistle Dedicatory" is a hymn to animals as symbolic images. He suggests that a history of beasts is preferable to a historical chronicle, because it reveals "that Chronicle which was made by God himselfe, every living beast being a word, every kind being a sentence, and al of them togither a large history, containing admirable knowledge & learning, which was, which is, which shall continue, (if not for ever) yet to the worlds end.'[35]

Jonston and the Demise of Emblematic Natural History

The dominance of natural history by similitude is so complete in the first half of the seventeenth century that one certainly expects Joannes Jonston's multivolume *Natural History* of 1650 – which looks for all the world like another Renaissance encyclopedia – to conform to the Aldrovandi model.[36] All the Aldrovandi illustrations are there, as well as those of Gesner and assorted other Renaissance naturalists. It is a shock, then, to read the text of Jonston's work and realize that, with its publication, the bottom has suddenly dropped right out of the emblematic cosmos.

Joannes Jonston (1603–1675) is not well appreciated by historians of science. He is usually portrayed – when he is portrayed at all – as a second-hand Aldrovandi, and thus a thirdhand Gesner – the last of the Renaissance encyclopedists. It is hard to understand how this image of Jonston has persisted, for the text of his work reflects a remarkable metamorphosis. The entry on *pavo* can serve again to illustrate these changes.[37] It has been trimmed to a tidy two pages. There is a full description – nothing has been cut here – and a discussion of medical applications and culinary uses. But if one looks for peacock emblems, proverbs, or hieroglyphics, there are none to be found. Not a single reference to Camerarius, or Horapollo, or Erasmus – not in the peacock

34   The harshest criticism of Topsell came from Charles E. Raven, *English Naturalists from Neckham to Ray: A Study of the Making of the Modern World* (New York: Kraus Reprint, 1968), who called Topsell "unimaginative, commonplace. . . . He was not a man of high distinction, intellectual or practical," pp. 219–20.

35   Edward Topsell, *A Historie of Four-Footed Beastes* (London, 1607), sig. A5v.

36   Joannes Jonston, *Historia naturalis*, 6 vols. (Frankfurt, 1650–1653). The six volumes are on quadrupeds, birds, serpents, fish, marine invertebrates, and insects; all except the last were published in 1650.

37   Jonston, *Historia naturalis de avibus* pp. 56–8.

article, not in any article. Even the medicinal uses have been weeded out: The ones that suggest sympathetic cures are gone; those that allow a physical cause are retained.[38] In fact, Jonston's description of the peacock is virtually identical to that of Francis Willughby twenty-five years later. It is apparent that emblematic natural history began to wane long before the Royal Society took a dislike to it.[39]

It was Michel Foucault who suggested that Jonston's encyclopedia marked a clear break with earlier Renaissance natural history, and he does seem to have pointed his finger in the right direction, if not to the precise spot.[40] Something profound had indeed occurred around midcentury, and historians of other fields have noticed it, although they have placed the date of transition earlier or later. One description of the transformation, by François Jacob, is particularly eloquent:

> Living bodies were scraped clean, so to speak. They shook off their crust of analogies, resemblances and signs, to appear in all the nakedness of their true outer shape. . . . What was read or related no longer carried the weight of what was seen. . . . What counted was not so much the code used by God for creating nature as that sought by man for understanding it.[41]

Historians of linguistics have called this metamorphosis the "decontextualization" of the world; historians of magic the "disenchantment" or "desymbolization" of nature.[42] Historians of the natural sciences have simply not noticed it. But Foucault is right; Jonston's natural history is indeed a watershed publication. To Foucault, however, the "event" of Jonston's work is an enigma, one of those transitions that cannot be explained. In truth, there are some explanations for the sudden death of "animal semantics," to use Foucault's own evocative term. I would like to offer several here.

New World Natural Histories

Certainly one important factor in this mild revolution was the appearance, in the early decades of the seventeenth century, of the first natural histories of

---

38   Our other specimen animal, the hedgehog, has an entry in the volume *De quadrupetibus*, pp. 170–1. The entry is about one-tenth the size of Aldrovandi's article.

39   Willughby's *Ornithologia* (London, 1676) is often referred to as an example of the "new" natural history inspired by the Royal Society.

40   Foucault, *Order of Things*, pp. 128–130 (citing Jonston *Historia naturalis*, 1:1) [. . .]

41   François Jacob, *The Logic of Life: A History of Heredity*, trans, Betty E. Spillman (New York: Vintage Books, 1976), pp. 28–9. [. . .]

42   M. M. Slaughter, *Universal Languages and Scientific Taxonomy in the Seventeenth Century* (New York: Cambridge University Press, 1982), pp. 56–7; Peter Fingesten, *The Eclipse of Symbolism* (Columbia: University of South Carolina Press, 1970), p. 54. See also Owen Hannaway, *The Chemists and the Word: The Didactic Origins of Chemistry* (Baltimore: Johns Hopkins University Press, 1975), who does not give the transformation a name but describes it beautifully.

New World animals: Charles L'Ecluse's *Exotica* (1605), Jan de Laet's *New World* (1625), Juan Nieremberg's *History of Nature* (1635), and most important, the *Natural History of Brazil* (1648) by Georg Markgraf (1610–1644).[43] These natural histories are occasionally brought into survey accounts of the Scientific Revolution, but their significance is usually seen to lie in their demonstration of a Baconian explosion of knowledge. This, of course, is true, but New World narratives had a far greater influence than simply enlarging the subject matter of natural history. Their impact derived from one simple fact: The animals of the new world had no known similitudes. Anteaters and sloths do not appear in Erasmus or Alciati or Piero Valeriano; they are missing from all the writings of antiquity. They came to the Old World naked, without emblematic significance. Thus naturalists could not approach this new fauna in the manner of Aldrovandi. Instead, they were forced to limit their descriptions to discussions of appearance, habitat, food, and whatever tales could be assembled from native populations. The tension between Old World and New World natural history is particularly evident in the narrative of Juan Nieremberg (1595–1658). He begins his work with a sixteen-page first chapter that is a masterful – indeed rhapsodic – restatement of the emblematic view of nature.[44] Then he parades by the reader a host of capybaras, marmosets, and pacas, and not a single one has a known similitude or emblematic meaning. All he can provide is a physical description and a picture. The contrast between a page of Nieremberg and a page of Aldrovandi is remarkable.

Jonston compiled his natural history from both kinds of sources: Aldrovandi and Gesner on the one hand, Nieremberg and Markgraf on the other. He was confronted – really the first to be so confronted – with this great incongruity of style: Old World animals, clothed in similitudes; New World animals, bereft of associations. Perhaps for uniformity, perhaps for personal perference, perhaps because he did not feel able to create an emblematic New World out of whole cloth, Jonston adopted the model of the New World description. The Old World animals lay naked to the observer's eye for the first time. And never again would they resume their emblematic garb.

Browne and the Quest for Truth in Natural History

There were other factors involved, however, in the demise of the emblematic world view, for we can also see it under attack in a work radically different from

---

43 Charles L'Ecluse, *Exoticorum libri decem* (Leiden, 1605); Jan de Laet, *Novis orbis* (Leiden, 1633); Juan Eusebius Nieremberg, *Historia naturae* (Antwerp, 1635); Georg Markgraf, *Historia naturalis Brasiliensis* (Leiden, 1648). [. . .]
44 Nieremberg, *Historia naturae*, pp. 1–16

Jonston's *Natural History*, namely the *Pseudodoxia epidemica* (1646) of Thomas Browne (1605–1682) or, as it is sometimes called, the *Vulgar Errors*.[45] The *Pseudodoxia* is a concerted attempt to purge natural history of commonly, but erroneously, perceived truths. Many people believed that the badger has legs that are shorter on one side than the other; that the chameleon subsists on air and the salamander survives in fire; that a dead kingfisher, hung by the bill, will point in the direction of the wind. Such ascriptions, and hundreds more, can readily be found in the tomes of Gesner, Aldrovandi, and Topsell. But in the *Pseudodoxia*, Browne asks the remarkable questions: Are these stories true? Can they be demonstrated? By appealing to a threefold criterion of reason, experiment, and authority, Browne proceeds to evaluate a large number of such Vulgar Truths. Can a dead kingfisher truly function as a weathervane? Browne hangs several birds outside and finds that no two point in the same direction. Do toads and spiders have a mutual, innate antipathy? Browne decides the matter by placing a toad and several spiders in a jar, and he relates that the spiders crawled all over the unperturbed toad, who swallowed them contentedly, one by one, as they came near his mouth.[46]

Interestingly, in view of our specimen bird, Browne even puts the peacock to the test. Two of Aldrovandi's statements attracted Browne's notice: that peacocks are ashamed of their own feet, and that cooked peacock meat does not spoil.[47] Concerning the first, Browne says that the notion probably arose because the peacock must keep its head back to maintain its display of feathers; if the head inclines forward, the train collapses. It is not a matter of shame but of mechanics. Browne also does an experiment to test the purported non-putrefaction of roasted peacock flesh and discovers it to be true. But, as he points out, it is also true of the meat of many fowl – turkey and pheasant, for example – and so it is hardly a special virtue of the peacock.

Browne clearly has a different view of nature from Aldrovandi; he is uninterested in aphorisms or emblems that are not *true*. His skepticism is even more remarkable when we note that Browne was a true romantic (if the term makes sense when applied to the English baroque), a writer whose most famous sentiment was "I love to lose my self in a mystery, to pursue my Reason to an *O altitudo*."[48] Where did such a man acquire the idea that natural history involves

45   Thomas Browne, *Pseudodoxia epidemica: Or, Enquiries into very many received Tenents, and common presumed truths* (London, 1646).

46   Browne, *Pseudodoxia epidemica*, pp. 175 (toad), 157–63 (chameleon), 138–40 (salamander), 127–9 (kingfisher), 115 (badger). Two of the best studies of the *Pseudodoxia* are by Robert R. Cawley: "The Timeliness of *Pseudodoxia epidemica*" and "Sir Thomas Browne and His Reading." Both are in *Studies in Sir Thomas Browne*, ed. Robert R. Cawley and George Yost (Eugene: University of Oregon Books, 1965), pp. 1–40, 104–66.

47   Browne, *Pseudodoxia epidemica*, pp. 172–3.

48   Thomas Browne, *Religio medici*, in *Works*, 6 vols, ed. Geoffrey Keynes (London: Faber & Gwyer, 1928–1931), 1:13.

the separation of the true from the false? He did not arrive at these views by reading New World natural histories. I would like to suggest that the inspiration came from seventeenth-century antiquarianism.

Antiquarianism and the Quest for Historical Truth

Antiquarian studies changed markedly in the early seventeenth century, as part of what has been called, perhaps overenthusiastically, a "historical revolution."[49] The antiquarianism of sixteenth-century Italy was not considered a historical discipline. As Arnaldo Momigliano pointed out in a now-classic essay, antiquities in Italy were not used as the tools of history, because the history of ancient Rome and Greece had already been written – by Livy and Caesar and Polybius.[50] Thus the coins and relics unearthed in such abundance were put to other uses; they were mined for their emblematic value, as we have already seen, or they were simply amassed in collections, in the museums of Francesco Calzolari, Ferrante Imperato, Michele Mercati, and Aldrovandi.[51] In very few instances in the sixteenth century do we find a historian treating a coin or burial urn as a piece of historical evidence to be used in reconstructing the past.

But antiquarianism began to take quite a different turn in the northern countries around the end of the sixteenth century. Antiquarians in England and Denmark, in particular, began to see their artifacts as vital historical clues. The reason for the different attitude in the north is straightforwad: Northern countries had no classical, canonical histories.[52] Except for brief mentions in Caesar and Tacitus, the ancient history of England was a blank. There were, of course, medieval histories that purported to take England back to its first "plantation" – the works of Geoffrey of Monmouth, Gildas, and Bede – but as their authenticity came to be challenged in the late sixteenth century, the void began to be filled with reconstructions based on artifactual evidence.

In England we see this quite clearly in the work of William Camden (1551–1623), whose *Britannia* of 1586 was a prodigious attempt to reconstruct the entire face of Roman Britain from such things as coins, inscriptions, and the remains of Roman roads.[53] The artifact was being given a new power,

49   F. Smith Fussner, *The Historical Revolution: English Historical Writing and Thought, 1580–1640* (New York: Columbia University Press, 1962). For one reaction, see Joseph H. Preston, "Was There an Historical Revolution?", *Journal of the History of Ideas*, 38 (1977):353–64.
50   Momigliano, "Ancient History and the Antiquarian."
51   On the nature of the sixteenth-century museum, see the collection of essays in Oliver Impey and Arthur MacGregor, eds, *The Origins of Museums: The Cabinet of Curiosities in Sixteenth- and Seventeenth-century Europe* (Oxford: Oxford University Press [Clarendon Press], 1985).
52   Momigliano, "Ancient History and the Antiquarian," p. 7.
53   On William Camden, see Stuart Piggott, "William Camden and the 'Britannia,'" in *Ruins in a Landscape: Essays in Antiquarianism* (Edinburgh: Edinburgh University Press, 1976), pp. 33–53;

and the antiquaries were consciously aware of it. Camden declares, for example, that you can learn more about medieval dress from monuments, glass windows, and reliefs than from the writers of those times.[54] The artifact does not lie. It is this obsession with truth that really distinguishes post-Camden antiquarians from earlier collectors of antiquities and from literary historians. Camden says in the preface of his history of Queen Elizabeth's reign: "For the love of truth, as it hath beene the only spurre unto me to undertake this work; so hath it also been my onely scope and aime."[55] When we bear in mind that truth was not high on the list of the essential qualities of literary history – certainly it ranked below moral education as a virtue – we see what a revolution the artifact has wrought.

Antiquarianism and Natural History

Antiquarian history did not have an immediate impact on literary history. Bacon kept "Antiquarianism" and "Perfect History" quite separate in the *Novum organum*, and they remained apart until after the middle of the seventeenth century.[56] But the antiquarian spirit did have a considerable effect on natural history, because the two fields overlapped considerably. There was, after all, no firm line between the Saxon urn, the stone axhead, the fossilized shark tooth, the unicorn horn, and the agate. Most of the great museum collections of the first half of the century – those of Basil Besler of Nuremberg, Ole Worm of Copenhagen, or the Habsburg emperors in Prague and Vienna – contained a mixture of natural and antiquarian artifacts.[57] And so natural historians who were exposed to the antiquarian attitude toward evidence came to see the natural world quite differently from Aldrovandi.[58]

T. D. Kendrick, *British Antiquity* (London: Methuen, 1970), pp. 143–59; Hugh Trevor-Roper, "Queen Elizabeth's First Historian: William Camden," in *Renaissance Essays* (Chicago: University of Chicago Press, 1985), pp. 121–48; F. J. Levy, *Tudor Historical Thought* (San Marino, Calif.: Huntington Library, 1967), pp. 148–63; Fussner, *Historical Revolution*, pp. 230–52.

54   Camden, quoted in Piggott, "William Camden and the 'Britannia,'" p. 37.

55   William Camden, *The Historie of the Most Renowned and Victorious Princesse Elizabeth* (London, 1630), "To the Reader," sig. B1v; partially quoted (from a different translation) in Herschel Baker, *The Race of Time: Three Lectures on Renaissance Historiography* (Toronto: University of Toronto Press, 1967), p. 20.

56   J. G. A. Pocock, *The Ancient Constitution and Feudal Law: A Study of English Historical Thought in the Seventeenth Century* (Cambridge: Cambridge University Press, 1957: reprint edition, New York: Norton, 1967), pp. 6–7.

57   On early seventeenth-century museums, see Impey and MacGregor, *Origins of Museums*.

58   There is a good discussion of the interaction between natural history and antiquarianism (as well as other historical disciplines) in Barbara Shapiro, "History and Natural History in Sixteenth- and Seventeenth-century England: An Essay on the Relationship between Humanism and Science," in *English Scientific Virtuosi in the Sixteenth and Seventeenth Centuries* (Los Angeles: William Andrews Clark Memorial Library, 1979), pp. 1–55. See also my earlier dissertation, "The

Thomas Browne certainly falls into this category. He had a passionate inter-
est in antiquities. One of his finest prosodic rhapsodies, the *Hydrotaphia*, or
Urn-Burial, was inspired by the discovery of several Saxon burial urns in a
Norfolk tomb, and in many of his other writings and letters Browne manifested
a great fondness for the artifacts of the past.[59] All his works reflect an intimate
familiarity with the antiquarian scholars of his century: Camden, Worm,
John Twyne, John Stow, Richard Verstegan, Jan Goropius Becanus, William
Dugdale, and many more. He was much impressed by the ability of the antiquarian to wrest a truth from "the ruins of forgotten time" on the basis of
slight, but incontrovertible, evidence.[60] Browne tested Roman artifacts for
residual magnetism, attempted to determine the age and sex of exhumed
skeletons, and suggested how barrows could be dated by the presence of
"distinguishing substances."[61] In other words, he made artifactual evidence
the standard for determining historical truth, and he tended to ignore or down-
play the evidence of literary history, in spite of his own literary inclinations. It
is not surprising, then, that when Browne approached the writing of natural
history, he subjected the literary tradition there to the test of empirical evidence. And with this new conception of what constitutes natural history, the
entire emblematic tradition fell apart – or, more accurately, became irrelevant.
For Browne, animal symbolism was no longer a part of the study of nature,
because it had no basis in truth.

It would seem, then, that Thomas Browne and Joannes Jonston reformu-
lated natural history for quite different reasons but with rather similar results,
and, most interestingly to note, at almost exactly the same time.[62] But this is
still not the whole story. There is a third factor that should at least be considered in the decline of the emblematic world view, and that is Baconianism.
Several observers have pointed to Bacon as being an instrumental force in the
rise of a new natural history in the latter part of the seventeenth century,[63]

Sense of the Past in English Scientific Thought of the Early Seventeenth Century: The Impact of
the Historical Revolution," University of Wisconsin at Madison, 1975, which covers similar
ground.
59   Browne's *Hydrotaphia – Urne-Burial, or, A Brief Discourse of the Sepulchrall Urnes lately found
in Norfolk* was originally published in 1658; it is found in *Works*, ed. Keynes, 4:7–50.
60   A splendid example of an antiquarian deduction was John Stow's claim that the Romans
buried at least some of their dead in coffins, a claim buttressed solely, but powerfully, by the discovery of tiny nailheads set in a coffin-shaped array around many graves in a Roman cemetery
at Spitalfields. Browne refers to Stow and Spitalfields in the *Hydrotaphia* (*Works*, 4:17). On Spitalfields, see M. C. W. Hunter, "The Royal Society and the Origins of British Archaeology," *Antiquity*,
65 (1971):113–21, 187–92; p. 118.
61   Browne, *Hydrotaphia*, in *Works*, 4:18, 26; "Of Artificial Hills, Mounts, or Burrows," *Miscel-
lany Tracts*, in *Works*, 5:99–103, p. 102.
62   This conjunction of "discontinuities" would no doubt have pleased Foucault, although, were
he still with us, he would doubtless reject my attempts to give the transition a causal explanation.
63   See especially Joseph M. Levine, "Natural History and the History of the Scientific Revolution," *Clio*, 13 (1983):57–73. [. . .]

and it is not unreasonable to suppose that Bacon's views might also have been felt in the earlier age of Browne, Markgraf, and Jonston.

Bacon and the Real Language of Nature

Bacon never wrote a natural history; his posthumous *Sylva sylvarum*, of 1627, which is often called his "natural history," is in reality a heterogeneous collection of random observations and suggestions for further inquiry. But Bacon did have definite ideas on how a proper natural history should be written, and he thought that the existing natural histories were unsatisfactory, because, as his executor William Rawley put it, they showed the world as men made it, not as God made it; Bacon's natural history, in contrast, would have "nothing of Imagination" in it. And Rawley elaborated:

> For those Natural Histories which are Extant, being gathered for Delight and Use, are full of pleasant Descriptions and Pictures; and affect and seek after Admiration, Rarities, and Secrets. But contrariwise, the Scope which his Lordship intendeth is to write such a Naturall History, as may be Fundamental to the Erecting and Building of a true Philosophy; For the illumination of the Understanding; the extracting of Axiomes, and the producing of many Noble Workes, and Effects.[64]

What makes Bacon particularly striking, however, is that he not only spurned the use of the emblematic tradition in natural history; he rejected the entire emblematic world view as invalid. There is no web of correspondences for Bacon; similitudes do not lead to understanding; the universe is not written in a code that reveals the attributes of God.[65] Bacon was one of the first natural philosophers to take this stance. As early as the "Valerius terminus" of around 1603, Bacon had stated: "For if any man shall think by view and inquiry into these sensible and material things, to attain to any light for the revealing of the nature or will of God, he shall dangerously abuse himself."[66] And in his later writings Bacon regularly warns against trying to impose patterns on nature that do not really exist in nature. "There is a great difference," Bacon says, in aphorism 23 of his *Novum organum*, "between the Idols of the human

64   William Rawley, "To the Reader," in Francis Bacon, *Sylva sylvarum* (London, 1627), sig. A3r, A1v; or James Spedding et al., eds, *The Works of Francis Bacon*, 14 vols. (London: Longmans, 1857–1874), 2:335–7.

65   Paolo Rossi, "Hermeticism, Rationality, and the Scientific Revolution," in M. L. Righini Bonelli and William R. Shea, eds, *Reason, Experiment, and Mysticism in the Scientific Revolution* (New York: Science History, 1975), pp. 247–73; citing pp. 258–9.

66   Bacon, *Valerius terminus*, in *Works*, 3:218.

mind and the Ideas of the divine. That is to say, between certain empty dogmas, and the true signatures and marks set upon the works of creation as they are found in nature."[67] And elsewhere, more flatly: "The world is not the image of God."[68]

Bacon's rejection of the notion that the natural world is a divine language, encoded by God, is almost certainly related to his views on human language. The prevalent, Platonic tendency of the late Renaissance, as we have seen, was to consider the meanings of words as inherent in the words themselves, just as the meanings of animals lay embedded in their very natures. Words and things were all of a piece, and the entire world of objects, letters, signs, and symbols was part of one language, the meaning of which was built in by God.

Bacon argued for separating words from things. Words are not intrinsically connected to objects but are arbitrary and conventional. Their only meanings are the ones we assign to them.[69] Such a view of language, which ultimately (and ironically, considering Bacon's reputation) is derived from Aristotle, undermines to a considerable extent the emblematic world view. If words have no hidden meanings, why should nature? If the language of man is arbitrary, can there be a language of nature at all? How can the Book of Nature shed light on God's plan, if the language of that book is devoid of meaning? Bacon seems to have realized the implications and to have decided that nature is not a multilayered complex of signs and hieroglyphics and that philosophers need not concern themselves with such matters.

The Impact of Baconianism on Natural History

Baconianism thus contained the seeds of insurrection against the emblematic world view. But did these seeds bear immediate fruit? Did Baconianism play any role in the demise of that view? It seems that the answer is no. Thomas Browne was indebted to Bacon in various ways, especially in the importance he ascribed to experiment and observation, but Browne's view of nature seems independently arrived at and, in any event, is not especially Baconian. Jonston was not touched by Bacon at all, nor were the New World naturalists on whose work Jonston relied, such as Markgraf. In truth, Bacon's attitude seems to have had little impact on naturalists before the era of the Royal Society. If his presence was felt before then, it was so subtle as to be, shall we say, occult.

---

67 Bacon, *Novum organum*, in *Works*, 4:51.
68 Bacon, *De augmentis*, in *Works*, 4:341.
69 Martin Elsky, "Bacon's Hieroglyphs and the Separation of Words and Things," *Philological Quarterly*, 63 (1984):449–60.

Natural History, Antiquarianism, and the Scientific Revolution

We must conclude then that the dismantling of the emblematic world view was an event prior to, and independent of, the rise of Baconianism. It was also prior to, and independent of, the spread of Cartesian mechanism. Consequently, we historians might well rethink some of *our* commonly perceived truths about the relationship between the rise of the mechanical philosophy and the decline of the world of magic. We seem to take it for granted that the former caused the latter; that nature was stripped of its correspondences and occult forces by a generation of Cartesians committed to a philosophy that allowed only explanations grounded on matter in motion. In truth, Browne, Jonston, and their generation dispensed with sympathies and correspondences for entirely different reasons, because of developments outside the physical sciences, and even outside science itself.

One final point seems worth stressing. We have squeezed antiquarianism in through the back door here, by demonstrating its impact on natural history. But the influence of antiquarianism, and of seventeenth-century historical thought in general, is broader than this, and the interplay of science and history is one of the most neglected facets of seventeenth-century studies. The Scientific Revolution was, after all, itself a historical revolution. It changed forever the way we would view Aristotle, Ptolemy, Galen. It altered the very concept of historical process. It is no simple coincidence that scientists of the seventeenth century developed keen interests in such matters as the origins of language, the early geological history of the earth, the settlement of the New World, the chronology of Egyptian and Chinese history, the collection of fossils, the early history of Christianity. The union of antiquarianism with literary history fashioned by historians was very similar to the approach of natural philosophers who forged a workable alliance between experiment and authority. Both groups developed, really for the first time, a true historical sense, which allowed them to place the past in proper perspective and, consequently, opened up the possibilities of the present and future. I merely suggest here that the similarities are perhaps not coincidental. It may well be that the historical revolution played a greater role than we now appreciate in the reconstruction of world views that we call the Scientific Revolution.

6

Medicine and Alchemy

The Chemical Philosophy and the Scientific Revolution

Allen G. Debus

Originally appeared as "The Chemical Philosophy and the Scientific Revolution," in *Revolution in Science: Their Meaning and Relevance*, edited by William R. Shea (Canton, Science History Publications/USA, a division of Watson Publishing International, 1988): 27–48.

Editor's Introduction

Like Ashworth, Allen G. Debus advocates expanding the narrative of the Scientific Revolution to include other disciplines, ones which were not mathematical, quantitative, or mechanical, such as medicine and chemistry. Part of the shift away from teleological or Whig histories of science is that disciplines which scholars previously condemned as pseudo-science, such as astrology and alchemy, are now being subjected to serious study.

The history of science has not on the whole been very good at incorporating the history of medicine. Historians of medicine and of science have different educations, professional journals, and conferences. The result is that the early modern sciences and early modern medicine tend to be treated in separate histories implying that pure science and medicine are completely distinct activities. Debus shows us how inappropriate this is for the early modern period.

The case-study that Debus examines here is the contested process by which chemical medicine penetrated the medical establishment, such as the Paris medical faculty and the London College of Physicians, in the face of considerable opposition. Certainly the struggle was one in which physiological and etiological theory played a major role; in contrast to traditional Galenic practitioners, who believed disease was caused by an imbalance in the body's humors and in general could be cured by a proper regimen of habits and diet, chemical doctors who followed Paracelsus regarded disease as the result of an external agent which had to be treated with chemically derived medicines. But it was also a struggle for power and control over the medical market place. Religious issues played a role, too: in France the practitioners of the new chemical medicine were largely Protestant; those in the establishment, Catholic.

Debus prompts us to ask questions about what constitutes a revolution in science. Is it fundamental change in a discipline's central theories, what Thomas Kuhn has described as the replacement of one paradigm by another? Or is it change in the status of a discipline, for example, of mathematics overcoming its status as subordinate to natural philosophy? Or is it fundamental change in the site of a discipline, such as the acceptance of chemistry at the university? All of these changes occurred in various disciplines in the early modern period. Furthermore, similar developments can have markedly different trajectories in different places, as Debus's comparison of France, England, and Spain shows us. Also, he reminds us that a revolution need not produce the modern discipline in order to be revolutionary; seventeenth-century chemistry had made a decisive break with the ancients and achieved widespread acceptance, but it would require another revolution before it could be regarded as modern chemistry. Finally, we have here another revolution that has little or nothing to do with the standard narrative of the Scientific Revolution that focused on astronomy, mechanics, and mathematics.

The Chemical Philosophy and the Scientific Revolution

Allen G. Debus

The Scientific Revolution has been interpreted traditionally in terms of the changes that occurred in astronomy and the physics of motion in the century and a half that separates the *De revolutionibus* of Copernicus (1543) and the *Principia mathematica* of Newton (1687).[1] There is ample reason for this if we are to be guided primarily by the accomplishments of modern science. We are all aware that the discipline of the history of science has positivistic roots as it was developed in the work of George Sarton and many of his contemporaries. Most of these scholars were willing to dismiss vast areas of thought once important in the understanding of nature to the realm of "pseudo-science." So influential was their work that those who are concerned with the natural magic, alchemy, and astrology of the sixteenth and the seventeenth centuries must still turn to the section on the "pseudo-sciences" in the Critical Bibliography of *Isis*.

However, the research of the past three decades has made it increasingly evident that the history of physics and astronomy cannot be equated with the history of science as a whole. We are becoming ever more a ware of the importance of societal factors and earlier positivistic interpretations are gradually yielding to the demands of contextual history, which has blurred formerly distinct divisions between the sciences and the so-called pseudo-sciences. Another distinction made by Sarton between the history of science and the history of medicine has proved more difficult to eliminate. The debate between Sarton and Sigerist on this score is far from dead.

If we are willing to turn from the familiar script, that is, from Copernicus to Galileo to Newton . . . and if we are willing to view the biological sciences in terms other than the anatomical advance from Vesalius to Harvey, we rapidly become aware of other debates of pressing concern to sixteenth- and seventeenth-century scholars. Above all, we are struck by the fact that the scientific *and* medical literature of the period contains no more heated a debate than that centered on the innovations of Paracelsus (1493–1541) and his follower who called themselves Chemical Philosophers.

---

1 I have discussed the historiography of the history of science and the history of medicine in a set of four lectures presented at the University of Coimbra in 1984 and subsequently published as *Science and History: A Chemist's Appraisal* (Coimbra: Edição do Serviço de Documentação e Publicações da Universidade de Coimbra subsidiado pelo Fundação Calouste Gulbenkian, 1984).

The Paracelsian Chemical Philosophy[2]

In a period characterized by the search for the true texts of the ancients, the Paracelsians called for a new philosophy based on fresh observations . . . one that would be based on a truly Christian interpretation of nature and man. This call, frequently strident in its intensity, was to pit these men against both he medical establishment and the natural philosophers of the schools. Debate was inevitable and it was to be joined in both theoretical and practical arenas.

The Paracelsians rejected Aristotle and Galen – even when they borrowed concepts from them . . . they rejected deductive logic as a guide to truth, and in doing so they rejected mathematics (because of its geometrical method of proof) and mathematical abstraction as a proper approach to understanding natural phenomena.[3] Not only was Aristotelian logic to be scrapped, but also the four elements which had served as the very basis of physics, chemistry, and medicine for nearly two millenia. Rejecting both humoral medicine and Scholastic philosophy, they sought true knowledge through religion and the Hermetic and Neo-Platonic texts of late antiquity. Knowledge of nature was properly to be understood as a knowledge of God's Creation and was in a sense a truly religious exercise. There were two books of divine revelation. The first, the *Bible*, was to be read; the second, Nature, was the book of Creation, a book which was to be probed both in the field and in the laboratory, where use of the chemists' fire would surely reveal hitherto unknown secrets. An inseparable link between these two books was to be found in Genesis, Chapter 1, which presented to man the order of Creation. And because chemistry and alchemy seemed to these men to be the prime example of what an observational science should be, it is not too surprising to find them picturing the Creator as a divine alchemist who had separated the beings and objects of the earth and the heavens from the unformed *prima materia* much as an alchemist might distill pure quintessence from an impure form of matter.

Attention to the Creation raised the question of the elements, and since fire could not be supported as an element on the basis of Genesis, the traditional element system was invalid. Jean Baptiste van Helmont (1579–1644) proceeded to argue that since this was true, neither the physics nor the medicine of the ancients, could continue to be upheld. For him it was necessary to destroy them both and to establish a new philosophy, one that would be

---

2   See Allen G. Debus, *The Chemical Philosophy: Paracelsian Science and Medicine in the Sixteenth and Seventeenth Centuries* (2 vols, New York: Science History Publications, 1977), I, 63–126.

3   Allen G. Debus, "Mathematics and Nature in the Chemical Texts of the Renaissance," *Ambix*, 15: 1–28, 211 (1968); Allen G. Debus, "Motion in the Chemical Texts of the Renaissance," *Isis*, 64: 4–17 (1973). These papers have been reprinted with the original page numbers in Allen G. Debus, *Chemistry, Alchemy and the New Philosophy 1550–1700* (London: Variorum, 1987).

grounded on fresh observations and Christian truths.[4] A century earlier Paracelsus had shopped short at discarding earth, water, air, and fire, but he had introduced the *tria prima*, sulphur, salt, and mercury, as a more useful basis for understanding matter. Thus, not only the direct questioning of the Aristotelian elements, but the introduction of a new system had the effect of calling into question traditional natural philosophy and medicine.

It is important to keep in mind that this Chemical Philosophy of the Paracelsians went far beyond the subject of medicine.[5] Our world had been created through a divine alchemy, and it was best understood to be continually operating in a chemical fashion. Meteorological phenomena were explained chemically as were volcanic action, the origin of mountain springs, and the generation and growth of metals and minerals. Even agriculture was seen as a fitting subject of investigation, and Paracelsus explained the beneficial action of manure by the presence of its life-giving salt. In short, nothing was to be excluded from the framework of this new philosophy.

Nevertheless, the overriding interest of the Paracelsians was to be found in medicine.[6] The Chemical Philosophy was universal to be sure, but its practitioners were for the most part physicians and their acceptance of the macrocosm-microcosm universe assured the macrocosmic investigator of information that would benefit his medical practice. Thunder and lightning had been explained by a reaction of an aerial sulphur and niter that was seen to be similar to the explosion of gunpowder. But surely the same aerial reaction could result in hot and burning diseases in the body since these aerial substances could be inhaled and react within the body. As for the humoral medicine of the ancients, it was to be rejected. In its place the Paracelsians spoke of localized diseases. The organs of the body were presided over by specific archei, which separated pure nourishment from gross substance much as an alchemist works in his laboratory. But an external agent of disease might enter the body through the air or through the digestive tract and lodge in a specific organ. Indeed, should the archeus of the disease entity overpower the archeus of that organ, death could result.

The late medieval and early modern periods had witnessed a number of terrifying diseases ranging from the depredation of the Black Plague to the

---

4 Debus, *Chemical Philosophy*, II, 295–379 on van Helmont. See also Walter Pagel, *Joan Baptista Van Helmont: Reformer of Science and Medicine* (Cambridge et al.: Cambridge University Press, 1982). In his "Promissa authoris," van Helmont stated that it was necessary to "destroy the whole natural Philosophy of the Antients, and to make new the Doctrines of the Schools of natural Phylosophy." John Baptista van Helmont, *Oriatrike or Physick Refined. The Common Errors therein Refuted, And the whole Art Reformed & Rectified. Being A New Rise and Progress of Phylosophy and Medecine, for the Destruction of Diseases and Prolongation of Life*, trans. J(ohn) C(handler) (London: Lodowick Loyd, 1662), I.
5 Debus, *Chemical Philosophy*, I, 84–96.
6 Ibid., I, 96–109.

venereal diseases. Those involved in the practice of medicine sought something beyond a theoretical Chemical Philosophy. They needed cures for life-threatening diseases. Here, too, chemistry was brought to play.[7] The diseases of the new age were pictured as far worse than those of antiquity and the argument was made that they required stronger medicines than those of the traditional herbalists.[8] Distillation would give the physician the pure essence of plant substances and this was presented as a more efficacious procedure than a simple mixture of dried plan remains. But more debatable at the time was the emphasis on inorganic chemicals for purging. The Galenists argued forcefully that many of these chemicals were poisons and should be avoided at any cost. Traditional medicine taught that diseases – as bodily poisons – were to be cured by their opposites.[9] The Paracelsians replied that like cures like, that is, that poisons could only be cured by poisons in proper measure. They claimed both that they had altered the poisonous nature of these substances chemically, so that they would not harm the body, and also, in contrast to the claims of the Galenists, that they paid close attention to dosage, so that no one need fear excessive quantities.

The Chemical Debates[10]

The scope of the developing debate was first hinted at in the multivolume *Disputationes de Medicina Nova Paracelsi* (1572–1574) of Thomas Erastus (1524–1583), the Professor of Theology and Moral Philosophy at Basel.[11] For Erastus, Paracelsus had been an evil magician and medical charlatan who had neglected logic and had not been properly educated in the truths of the ancients. It was blasphemous to compare the divine Creation to a chemical separation, while the macrocosm-microcosm analogy was nonsense. He rejected the three principles and insisted that the Aristotelian element system was one of the glories of antiquity.

7   Ibid., I, 112–117.
8   The Elizabethan chemist, John Hester, wrote that it was useless to turn the pages of Hippocrates and Galen searching for cures of diseases that had not existed in antiquity. Specifically they had not known about the new venereal diseases "and therefore as this latter age of ours sustaineth the scourge thereof, a iust whyp of our lycentiousness, so let it (if ther be any to be had) carry the credite of the cure, as some rewarde of some mens industries." This is from Hester's preface to Phillip Herman's *Treatise teaching howe to cure the French-Pockes . . . Drawen out of that learned Doctor and Prince of Phisitions, Theophrastus Paracelsus*, trans. John Hester (London, 1590), 31.
9   Walter Pagel, *Paracelsus: An Introduction to Philosophical Medicine in the Era of the Renaissance* (Basel: S. Karger, 1958), 141–50 on "aetiological and specific therapy."
10   Debus, *Chemical Philosophy*, I, 126–204.
11   Ibid., I, 131–134.

Erastus was a physician as well as a theologian. Here, too, he took issue with the Paracelsians. He insisted that the basis of true medicine was the humoral theory of the ancients, not the new doctrine that diseases are separate entities that enter man from without. But above all, Erastus took issue with the chemically prepared medicines, which had introduced all kinds of lethal metals and minerals for internal use.

This brief reference to Erastus indicates that the Paracelsian debates of the late sixteenth and the seventeenth centuries spread far beyond the simple question of the medical use of inorganic chemical medicines. To be sure this is an issue of great importance for our understanding of the confrontation, but it should not be forgotten that there were some authors who were far more concerned with the implications of the mystical Paracelsian cosmology and its relation to religion – and in time there were others who were more interested in the political overtones involved in a confrontation with the medical and educational establishments.

We could easily emphasize here the theoretical statements of authors such as Joseph Duchesne (Quercetanus) (c. 1544–1609) or perhaps the "Admonitory Preface" of Oswald Croll (c. 1560–1609) to his widely read *Basilica Chymica*. Even more to the point might be the debates of Robert Fludd (1574–1637) with Johannes Kepler (1571–1630) regarding the cosmological harmonies of the universe or the use of mathematics – or Fludd's later debates with Marin Mersenne (1588–1648) and Pierre Gassendi (1592–1655) regarding the place of alchemy in the study of nature and medicine. Or we might turn to the persistent call for the reform of education by the Chemical Philosophers who pictured the university system as the bulwark of Aristotelian natural philosophy and Galenic medicine. This was the view of Van Helmont no less than it was for his contemporaries Bacon and Descartes. As late as 1654 John Webster argued that only by a drastic curricular reform at Oxford and Cambridge based largely upon the Chemical Philosophy would students have the benefits of a truly Christian education.[12]

However, important as the theoretical debates relating to the chemical cosmology of the Paracelsians may have been, for many physicians a more immediate threat was seen in the chemical medicines whose use seemed a real threat both to the lives of their patients and to the bases of ancient medicine.

The course of these debates is then of special interest – and for several reasons. They are of interest for their place in the general development of the Scientific Revolution because of the confrontation they present between

12   John Webster's treatise on educational reform and the replies by Seth Ward, John Wilkins, and Thomas Hall have been reprinted in Allen G. Debus, *Science and Education in the Seventeenth Century. The Webster-Ward Debate* (London: Macdonald and Co., Ltd./New York: American Elsevier, 1970). The debates between Robert Fludd, Johannes Kepler, Marin Mersenne, and Pierre Gassendi are discussed in Debus, *Chemical Philosophy*, I, 256–279.

"ancients" and "chemists" no less than between "chemists" and mechanists. But they are of interest as well in indicating the spread of a new philosophy of nature and medicine in a crucial period of the rise of modern science and medicine. Here we might well center our discussion on Central Europe, since it was there that the origin of the Paracelsian heresy was to be found. However, we will turn instead to a comparison of the reception of the new Chemical Medicine in England, the Iberian Peninsula, and, above all, in France where it is possible to see differing reactions due to regional conditions which varied from one country to another. In short, the chemical debates emerge as an essential chapter in the Scientific Revolution as a whole, and I believe that the history of the gradual acceptance of chemistry by the academic community should be seen as an essential part of the Scientific Revolution.

Chemical Medicine in France 1560–1660[13]

Widespread knowledge of the work of Paracelsus in France seems to have developed first during the 1560s. A French translation of Pierre Mattioli's (1501–1577) commentary on Dioscorides – published first in 1544 in Latin – appeared in 1561. Here was to be found a chapter on antimony in the fifth book on stones, minerals, and metals.[14] In the discussion of antimony's medicinal use, it was stated that the substance was good for plague, melancholic humors, persistent fevers, asthma, spasm, and paralysis among other ailments. However, it was also noted that only one author, Paracelsus, had described its internal use. Drawing on this, George Handsch (1529–1578)[15] had discussed the use of antimony as a purge and explained its success by comparing it with the purification of gold. As antimony may be used to rid gold, the most perfect of metals, of its impurities, so too it purges man, God's most perfect living creation, in similar fashion.

Drawing upon this description in Mattioli, a physician from Montpellier, Louis de Launay (fl. 1557–1566), prepared a tract on the marvelous qualities of antimony (1564). He declared that antimony was not a poison and that he had taken it himself on numerous occasions without ill effects.[16] Two years

13 Debus, *Chemical Philosophy*, I, 145–173. I am presently completing a book on chemical medicine in France in the period 1500–1700.

14 M. Pierre Andre Matthioli, Medecin Senoys, *Les Commentaires . . . sur les six liures des Simples de Pedacius Dioscoride Anazarbeen* (Lyon: A L'Escu de Milan par Gabriel Cotier, 1561). Here the chapter on antimony will be found on pp. 444–45. The pagination is the same for the 1566 edition.

15 Georg H. Handsch (1529–1578) prepared the German translation of Mattioli's commentaries on Dioscorides: *New kreüterbuch mit den allerschönsten und artlichesten figuren aller gemechtz . . .* (Prague: G. Melantrich von Aventine und V. Valgriss, 1563).

16 Loys de Launay, Medecin ordinaire de la Rochelle, *De la Faculté & vertu admirable de l'Antimoine, auec response à certaines calomines* (La Rochelle: Barthelmi Berton, 1564), sig. Di[r].

later, Jacques Grévin (1538–1570) of the Medical Faculty at Paris replied to de Launay warning of the poisonous nature of antimony. Due to his influence the Parisian Faculty of Medicine declared antimony a poison on August 3, 1566, and this exchange launched a bitter debate that was not to be settled for a century.[17]

It is of interest to note that the debate began over the internal use of a single substance and that the exchange between de Launay and Grévin scarcely mentioned Paracelsus. This was to change rapidly. In a second exchange between the two authors the name Paracelsus figured more prominently, and when Grévin translated Jean Wier's (1515–1588) five books on devils, enchantments, and sorceries in 1567, he included a damning appraisal of Paracelsus.[18] In the same year Pierre Hassard (fl. 1566–1570) translated the *Greater Surgery* of Paracelsus into French and Jacques Gohory (Suavius) (1520–1576) published a short compendium of Paracelsian philosophy.[19] Both of these authors praised Paracelsus and his chemical cosmology at a time when Grévin was attacking him as an arrogant liar and claiming that chemistry at best could be considered a small part of medicine, of use only for the separation of essences from their crude outer husks.

Despite resolute opposition to the chemists by the Medical Faculty of Paris, interest in the work of Paracelsus and in the medical application of chemistry grew rapidly in the closing decades of the sixteenth century. The most prominent French proponent of the new remedies quickly came to be recognized as Joseph Duchesne who entered the pamphlet warfare in a debate with the Galenist, Jacques Aubert (d. 1586) in 1575 over the origin of metals and the use of chemical medicines. The following year he published his *Sclopetarius* in which he discussed the serious problem of gunshot wounds. Here he compared the cleanliness of chemically prepared medicines with the filth of the Galenic concoctions.[20] In the important question relating to the cause of

---

17   The text of this decree is presented by Jacques Grévin in *Le Second Discovers de Jacques Grévin, Docteur en Medecin à Paris, sur les vertus & facultez de l'Antimoine, Avqvel Il est sommairement traicté de la nature des Mineraux, venins, pestes, & de plusieurs autres questiõs naturelles & medicinales, pour la confirmation de l'aduis des Medecins de Paris, & pour seruir d'Apologie contre ce qu'a escrit M. Loïs de Launay, Empirique* (Paris: Iacques du Puys, 1567) in Latin (sig. vii^r) and in French (fol. 101^r)

18   Jean Wier, *Cing Livres de l'impostvre et tromperie des diables: des enchantements & sorcelleries . . .*, trans. Jacques Grévin (Paris: Iacques du Puys, 1567), forl. 105^v.

19   I had access to the second editions of both works: Paracelsus, *La grande, vraye et parfaicte chirurgie*, trans. M. Pierre Hassard d'Armentieres, medecin et chirurgien (Anvers: Guillaume Silvius, 1568); Leo Suavius [Jacques Gohory], *Theophrasti Paracelsi philosophiae et medicinae utriusque universae, compendium, ex optimi quibusque eius libris: cum sholijs in libros IIII eiusdem De vita longa, plenos mysteriorum, parabolorum, aenigmatum* (Basel: Per Petrum Pernam, 1568).

20   I used the English translation by John Hester: Joseph Duchesne, *The Sclopotarie of Iosephus Quercetanus, Phisition, or His booke containing the cure of wounds received by shot of Gunne or such like Engines of warre, whereunto is Added his Spericke antidotary of medicines against the aforesayd wounds*, trans. John Hester, practitioner in the said spagiricall Arte (London, 1590), 74.

infection in gunshot wounds, Duchesne argued that lead is wholesome to our nature and that gunpowder is frequently used internally as a medicine by soldiers.[21] Therefore, these infections cannot be caused by these inorganic substances.

In 1578 the Faculty felt itself threatened by the arrival in Paris of Roch le Baillif (fl. 1578–1580), a Paracelsian from Brittany, who had been appointed physician in ordinary to Henry III.[22] Because le Baillif had extended his practice beyond the Court, the Faculty referred to its historic control of medicine in the area surrounding Paris and ordered le Baillif to halt his practice and his public lectures. He was then brought to trial and expelled from Paris.

The verdict of 1578 was looked upon as a significant victory for the Galenists, but in the end little was accomplished. The publishers continued to bring out chemically oriented medical texts. Indeed, there was enough interest in Paracelsus himself for Claude Dariot (c. 1530–1594) to prepare a new translation of the *Great Surgery* in 1588.[23]

For the most part, physicians interested in the new medicine were Protestants, while the Galenists were Roman Catholic. There are exceptions, but not many. This becomes a point of significance in the final decade of the century. Henry of Navarre took Paris in 1593. Although he converted to Roman Catholicism, he had been a Protestant, and he had with him physicians who favored the chemical medicine of the Paracelsians. Many came from families that had left France at the time of the St. Bartholomew's Day Massacre of 1572.

In 1594 the new King appointed Jean Ribit (c. 1571–1605) as his first physician.[24] As one who favored chemical medicine himself, Ribit appointed both Joseph Duchesne and Theodore Turquet de Mayerne (1573–1655) as Court physicians. Duchesne was a Calvinist who had taken his degree at Basel and had served the Swiss Cantons for many years. The parents of Mayerne had left France at the time of the 1572 Massacre for Switzerland, where he was born in 1573. He was to take his degree at Montpellier in 1597, a medical school which by this time was associated with chemistry, a subject of concern to the Medical Faculty at Paris.

The situation was complicated by a rivalry between the Faculty and the medical faction at Court. The Faculty claimed the right to control medicine and

21   Ibid., 7–10.
22   Debus, *Chemical Philosphy*, 155–156.
23   This was reprinted a number of times with additional works on chemical medicine by Dariot. I used the third edition: *La grand chirurgie de Philippe Aoreole Paracelse . . . traduite en Francois de la version Latin de Iosquin d'Alhem . . . par M. Claude Dariot plus un discours de la goutle . . . Item III. Traittez de la preparation des medicaments* (Montbeliard: Iaques Foillet, 1608).
24   Hugh Trevor-Roper, "The Sieur de la Rivière, Paracelsian Physician of Henry IV," *Science, Medicine and Society in the Renaissance: Essays to Honor Walter Pagel*, Allen G. Debus, editor (2 vols, New York: Science History Publications, 1972), II, 227–250.

to restrict the licensing of physicians to its own graduates. While their claim encompassed the whole of France, in practice it meant the area around Paris. However, the Court physicians were understood to be beyond their control. At the end of the century, their numbers were increasing, while those of the Faculty were decreasing. And since the Court physicians included some of the most prominent chemical physicians, the stage was being set for a confrontation between the two groups.[25]

This confrontation, which was to expand to all parts of Europe, was ignited by the *Matter of the True Medicine of the Early Philosophers* published by Joseph Duchesne in 1603.[26] Duchesne's debate with Aubert more than a quarter century earlier had centered on practical aspects of chemical medicine. His *Le grand miroir du monde* published in the following decade had discussed the Creation epic, the elements, and the place of chemistry in the interpretation of nature. However, this text was presented as a lengthy poem, and many of its most important points appeared in the form of lengthy notes. Its readership may well have been primarily those interested in literature. However, it was the new work of 1603 that immediately caught the attention of the French medical establishment. Here, and in a work on the *True Hermetic Medicine*[27] published the following year, Duchesne defended chemistry theoretically and practically, both as a key to nature and as the font of the new chemical medicines. He rejected the traditional humors and called for the adoption of the three Paracelsian principles by everyone. Although he said that he could not be a disciple of Paracelsus himself, there is little doubt that his views were very similar to those of Paracelsus.

The works of Duchesne led to a debate that was watched closely by physicians in all parts of Europe. He was instantly attacked by senior members of the Medical Faculty of Paris, and an extensive polemical literature was soon in print. Andreas Libavius (1540–1616) found so much material that he was able to write a short history of the conflict as early as 1606[28] – and a much longer one the following year. Nor did the death of Duchesne in 1609 still the debate. Increasing interest in chemically prepared medicines led to a new decree from

---

25 I am indebted to Dr. Rio Howard for letting me refer to her unpublished paper on "Medicine and the Royal Patronage of Science in the Early Seventeenth Century" presented at the Toronto meeting of the History of Science Society, 1981.

26 Joseph Duchesne, *Liber De Priscorum Philosophorum verae medicinae materia, praeparationis modo, atque in curandis morbis, praestantia . . .* (Leipzig: Thom. Schürer and Barthol. Voight, 1613). I have discussed Duchesne's views as presented in his publications of 1603 and 1604 in my *Chemical Philosophy*, I, 160–167.

27 I used the following edition: Joseph Duchesne, *Ad Veritatem Hermeticae medicinae ex Hippocratis veterumáue decretis ac Therapeusi, nec non vivae rerum anatomiae exegesi, ipsiusáue naturae luce stabiliendam, adversus cuiusdam Anonymi phantasmata Respondio* (Frankfurt: Ex Officina Typographica Wolffgangi Richteri, Impensâ Conradi Nabenii, 1605).

28 Andreas Libavius, *Alchymia, recognita, emendata, et aucta, tum dogmatibus & experimentis nonullis . . . .* (Frankfurt: Excudebat Joannes Saurius, impensis Petri Kopffii, 1606).

the Medical Faculty forbidding the sale of any and all chemical remedies in France (1615).

Marin Mersenne was more concerned with the sweeping demands of some chemists that a mystical alchemical cosmology be established as the basis of a new philosophy.[29] Instead, in 1625, he called for a new science based on mathematics and said that chemists must be policed by an academy that would keep them from discussing religious, philosophical, and theological questions. For him chemistry was useful only for the preparation of some new medicines. It is clear in the debates between Fludd and Mersenne, Gassendi and Kepler that the mystical chemistry and cosmology described by some authors could be viewed as a true basis of a new philosophy. Surely Mersenne and Gassendi looked upon this interpretation of nature as a great danger to their own dream of a new science. For them the grandiose claims of these chemists were as dangerous as they seemed to be for the Galenists.

Although the Medical Faculty remained resolutely opposed both to the internal use of chemicals in medical practice and to all other applications of chemistry to medicine, they never seemed to be able to end the activities of their adversaries. Chemists had offered private courses in their subject in Paris from the early years of the century, and their many textbooks sold in considerable quantities. Guy de la Brosse (c. 1586–1641) drew up plans for a Botanical Garden, which was formally opened in 1640.[30] Included was a teaching program, and the first Professor of Chemistry was the Scot, William Davisson (c. 1593–c. 1669), who was appointed in 1648.

By this time, however, the chief Parisian champion of chemistry was Théophraste Renaudot (1584–1653), a Montpellier MD (1606) who had converted to Catholicism.[31] Deeply concerned with the problems of the poor, he had attracted the attention of Cardinal Richelieu at an early age. Moving from his country practice to Paris, he set up his Bureau d'Adresse about 1630. This Bureau included an employment agency; it offered a program for low interest

29   P. Marin Mersenne, *La verite des sciences. Contre les septiques ou Pyrrhoniens* (Paris: Toussainct Du Bray, 1625; reprint Stuttgart/Bad Cannstatt: Friedrich Fromann Verlag (Gunther Holzboog), 1969). I have discussed Mersenne's recommendations in my *Chemical Philosophy*, I, 265–266.
30   On Guy de la Brosse, see the work by Rio Howard referred to in note 25 as well as her "Guy de la Brosse and the Jardin des Plantes in Paris," in *The Analytic Spirit*, ed. Harry Woolf (Ithaca: Cornell University Press, 1981), *La Bibliothéque et le Laboratoire de Guy de la Brosse au Jardin des Plantes à Paris* (Geneva: Librairie Droz, 1983), and "Guy de la Brosse: Botanique et chimie au début de la révolution scientifique," *Revue d'Histoire des Sciences* (1978). In the last case I have worked from a photocopy of the galley proofs kindly sent to me by the author. See also Henry Guerlac, "Guy de la Brosse and the French Paracelsians" in *Science, Medicine and Society in the Renaissance*, ed. Debus, I, 177–199.
31   On Renaudot see Howard M. Solomon, *Public Welfare, Science, and Propaganda in Seventeenth Century France: The Innovations of Théophraste Renaudot*. Princeton: Princeton U.P., 1972). Also useful for the action against Renaudot is *Pascal Pupoul, La querelle de l'Antimoine (Essai historique)* (Thèse pour l'Doctorat en Médecine, Paris: Librairie Louis Arnette, 1928).

loans and free medical consultations; and it sponsored weekly conferences on every conceivable sort of medical, scientific, and ethical subject.

Although Richelieu gave strong support to Renaudot's activities, he was viewed with alarm by the Medical Faculty. To them he seemed to be establishing a second medical school in the city, and his work conflicted with their claim to control the medical profession in France. And if this were not enough, Renaudot had graduated from Montpellier and was dispensing chemicals in his clinic. When he planned to enlarge his facilities in 1640, the Faculty advanced charges against him citing the historic rights of the Parisian Faculty, his use of chemical medicines, and the medical heresies of the school at Montpellier. His two sons, medical students in Paris, were refused degrees because of the "grave injuries of their father," and even Richelieu could not prevail upon the members of the Faculty to cease their attack. The Cardinal's death in 1642 and that of the King the following year left Renaudot defenseless. He was ordered to cease all of his activities at the Bureau d'Adresse in December 1643 and stripped of nearly all his privileges.[32] It was only five years after the humiliation of Théophraste that his sons, Isaac and Eusebius, were allowed to take their doctorates.

Once again the Medical Faculty seemed to have defeated the chemists. And yet, the Faculty's opposition was not to last much longer. Already in 1638 they had published an official pharmacopoeia – one that included antimony as a purgative, to the great distress of the more conservative members. And in the following years, ever more members of the Faculty openly favored the medical use of this substance. In 1652 over 60 members of the Medical Faculty signed a document favoring the internal use of antimony.[33] In 1658 Louis XIV was cured in the field with an antimony purge.[34] Gui Patin (1601–1672), who had been the most inveterate enemy of Renaudot, could still write in 1655 that even the Koran was less dangerous than the works of Paracelsus,[35] but the hard core Galenist membership was declining with age, and when the Medical Faculty met in full session to determine the fate of the medical use of antimony in 1666, 92 of the 102 present voted in favor of placing emetic wine on the list of purgatives.[36] It was exactly one hundred years after the Faculty had first damned antimony as a poison.

32   Pilpoul, 182–189.
33   Reprinted in Eusèbe Renaudot, Conseiller Medecin du Roy, Docteur Regent en la Faculté de Medecin à Paris, L'Antimoine Ivstifie et L'Antimoine Triomphant ov Discours Apologetique faisant voir que la Poudre; & la Vin Emetique & les autres remedes souuerains pour guerir la pluspart des maladies, qui y sont exactement expliquées. Auec leurs preparations les plus curieuses tant de la Pharmacie, que de la Chymie (Paris: Iean Henault [1653]), sig. é ii^r.
34   Pilpoul, op. cit., 77; Solomon, op. cit., 220.
35   Gui Patin, Lettres . . . Nouvelle Édition augmentée de lettres médites, precedee d'une notice Biographique . . . par J.-H. Reveillé-Parise (3 vols, Paris: J.-B. Baillière, 1846), III, 47.
36   Pilpoul, op. cit., 84–85; Solomon, op. cit., 220.

Admittedly a very complex story, the acceptance of chemistry in France has a broader significance for the history of science and medicine. From the beginning the interest in chemistry was tied to medicine, specifically the internal use of antimony or its compounds as a purgative. And yet, the connection with Paracelsus expanded the debate to include the question of the relation of chemistry to the explanation of the entire microcosm-macocosm cosmology. An acceptance of the universal nature of chemistry by Duchesne and his colleagues early in the seventeenth century led to a debate that touched every corner of Europe and was to cause Marin Mersenne to point to the claims of the more mystical chemists as dangerous to those who sought a new basis to replace the works of Aristotle and Galen. The medical establishment in Paris was even more concerned for they looked on the chemists as dangerous innovators who not only sought to overturn the works of the ancient physicians, but also represented the rival school of Montpellier and unorthodox religious views. Their various battles with the chemists culminated in their successful action against Renaudot, but in the end the Faculty was to accept the new chemicals.

Chemical Medicine in England and Spain 1560–1700

The acceptance of chemical medicines in England was far less stormy than in France. To be sure, the London College of Physicians at first demanded that Fellows adhere strictly to Galenic teaching, but from an early date there was a willingness to accept chemically prepared remedies.[37] As in France, the first references to Paracelsus appear in books printed in the 1560s, but relatively few details of medicinal chemicals appeared until the next decade. Influential surgeons, including George Baker (1540–1600) in 1574 and John Banister (1540–1610) in 1589, praised some of the new chemicals, but for the most part as salves for external use. The only extensive defence of the Paracelsian system in English in the sixteenth century was that of R. Bostocke (fl. 1585) whose *Difference between the auncient Phisicke . . . and the latter Phisicke* (1585) discussed the chemical Creation, the Paracelsian principles, and the macrocosm-microcosm universe as well as the medical aspects of the new medicine.[38] However, for Bostocke Paracelsus was not an innovator. Rather, he had revived ancient truths that could be connected with the knowledge known to Adam before the Fall. For him acceptance of the chemical medicine had religious overtones, and the work of Paracelsus could be compared with that of Luther and the other Protestant reformers in religion.

---

37   Allen G. Debus, *The English Paracelsians* (London: Oldbourne, 1965), see chapter 2, "The Elizabethan Compromise," 49–85.
38   Ibid., 57–65.

London practitioners of the closing decades of the sixteenth century were able to choose from a wide variety of chemically prepared remedies available from John Hester (d.c. 1593),[39] who also translated into English some of the more practical tracts of Leonardo Fioravanti (d. 1588), Joseph Duchesne, and Philip Hermann (fl. third quarter of the sixteenth century). Few were willing to condemn all of the chemists, and the renowned Elizabethan surgeon, William Clowes (c. 1540–1604), argued that there were both true and quack Paracelsians. Physicians should be willing to borrow valuable remedies from any source, whether Galenic or Paracelsian.[40]

Nor was the London College of Physicians as inflexible as its counterpart in Paris. Thomas Moffett (1553–1604), a prominent Fellow of the College, was a friend of the Danish Paracelsian Peter Severinus (1540–1603) and had received his degree at Basel in 1584 where Paracelsus had taught.[41] His defense of chemical medicines appeared in 1584 and was dedicated to Severinus. It was to be reprinted three times in the various editions of Zetzner's *Theatrum Chemicum* during the seventeenth century. It is little wonder that when the College planned the publication of an official pharmacopoeia, Moffett was placed in charge of the committee on chemical medicines (1589).

The English connection with Continental Paracelsism was not limited to Moffett's friendship with Severinus. Far more important was the French connection. John Hester had translated Duchesne's early works on gunshot wounds and on the use of chemical medicines. And a minister, Thomas Tymme (d. 1620), translated large sections of Duchesne's theoretical discussions of Paracelsian cosmology and medicine published in 1603 and 1604 as *The Practise of Chymicall, and Hermeticall Physicke* (1605). Like Bostocke, he saw in the Paracelsian system an approach to the study of nature and man that avoided the use of the "heathenish" texts of antiquity.

Knowledge of the French debate was not limited to hearsay or the translations of Hester and Tymme. Duchesne's ally in Paris in 1603 had been his colleague, Theodore Turquet de Mayerne (1573–1655), whose *Apologia* for Duchesne had earned him the hatred of the Medical Faculty.[42] Mayerne moved to London in 1606 where he was appointed physician to the Queen. After a brief return to Paris, he left his homeland permanently for England at the death of Henry IV in 1610. He was then appointed Chief Physician to James I and made a Fellow of the College of Physicians. Deeply interested in all aspects of chemical medicine, Mayerne joined a College that was by this time open to

39   Ibid., 66–69.
40   William Clowes, *A Right Frutefull and Approoved Treatise for the Artificiall Cure of that Malady called in Latin Struma* (London, 1602), see the "Epistle to the Reader."
41   Debus, *English Paracelsians*, 70–76.
42   On Mayerne's life, see Thomas Gibson, "A Sketch of the Career of Theodore Turquet de Mayerne," *Annals of Medical History*, New Series, 5: 315–326 (1933).

the consideration of all aspects of the subject. Even the alchemically inclined physician, Robert Fludd, had been accepted as a Fellow in 1607.

The earlier plan to issue a pharmacopoeia had by this time been shelved, but it was of great interest to Mayerne who was instrumental in reactivating the project. The *Pharmacopoeia Londinensis* did, in fact, appear in two editions in 1618, and in addition to the traditional Galenic preparations, there were sections on salts, metals, and minerals, on chemical oils, and on what was termed "more useful chemical preparations."[43] The preface, probably written by Mayerne, referred to the Fellows' veneration of traditional medical learning, and also to their willingness to use the newer chemical remedies.

In short, although influenced by the work of French authors, the English reaction to the chemical innovations of the sixteenth century was quite different from the developments in Paris. There, the Medical Faculty tried to halt the advance of chemistry, which was looked on as a threat to tradition. In contrast, in England there were proponents of the new medicine present in the College of Physicians at an early date, and there was relatively little resistance to the election of chemists during those same years, which witnessed the most vicious battles between the Medical Faculty and the Court physicians in France. However, England did not boast of two historically eminent medical schools of the level of Montpellier and Paris . . . nor was there the same underlying religious tension associated with the new medicine that caused friction in France. But in London as in Paris, the medical establishment sought to extend its control over the profession. Here they watched unaffiliated chemists with care. The 1617 Charter to the Apothecaries permitted them to dispense, but not prescribe medicines, and the 1618 *Pharmacopoeia* may be viewed as a guide to the preparation of medicines. Also in 1618, a Royal Charter to the College gave the Fellows permission to search the shops of apothecaries, distillers and other preparers of chemical medicines. And 20 years later the Fellows of the College were instrumental in establishing the City Company of Distillers. This, too, may be seen as an extension of their own rights of control. Indeed, by mid-century the College had appointed its own chemist and set up its own laboratory, so that chemical medicines could be obtained directly from them.[44]

The difficulties encountered by chemists in France differed considerably from the relatively easy acceptance of these remedies in England. But the situation in Spain and Portugal was unique, due to local conditions and religious

---

43   George Urdang, "How Chemicals Entered the Official Pharmacopoeias," *Arch. Int. Hist. Sci.*, 33: 304–313 (1954). See also the introduction to the *Pharmacopoeia Londinensis of 1618 Reproduced in Facsimile with a Historical Introduction by George Urdang* (Madison: University of Wisconsin Press, 1944).
44   Cecil Wall, H. Charles Cameron, and E. Ashworth Underwood, *A History of the Worshipful Society of Apothecaries of London*, I (London/New York/Toronto: Oxford University Press for the Wellcome Historical Medical Museum, 1963), 93.

restrictions.[45] Fear of the Protestant Reformation had led Philip II to attempt to separate Spain from foreign influences as early as 1557–1558, a date prior to the widespread dissemination of Paracelsian ideas. Both the Council of Castile and the Spanish Inquisition were charged with the licensing of books, and those who published or circulated unlicensed books were subject to death and confiscation of goods. In a similar fashion, higher education was closed to foreign influences. As part of the same draconian measures, Spanish students were ordered to return home from all foreign universities within four months, except for Bologna, Rome, Naples, and Coimbra. Needless to say, in the century from 1560 to 1660, the Spanish universities did not keep pace with others in Europe. This is particularly true of the sciences at a crucial time in the development of the Scientific Revolution. The cost of preserving religious orthodoxy was high.

But here we are concerned primarily with the place of chemistry in the larger fields of science and medicine.[46] In 1585 the Spanish *Index* took note of Paracelsus by expurgating several passages to be found in his lesser work on surgery. The edition of 1632 classified him as a Lutheran and forbade the reading of most of his work. Indeed, the only work that may be called Paracelsian in this period is a short work on practical chemical remedies published in 1589 that exists today in a single copy.[47] In the Iberian peninsula, medical instruction remained uncompromisingly Galenic throughout this period, which – as we have noted – was characterized by debate and open discussion in other parts of Europe.

A new interest in the sciences occurred only in the 1660s due to Don Juan of Austria who acted as Regent during the minority of Charles II.[48] He relaxed

45   A considerable amount of research over the past few decades has done much to uncover the period of the Scientific Revolution in Spain. Here see J.M. López Piñero, V. Navarro Brotóns and E. Portela Marco, "Selección bibliográfica de estudios sobre la ciencia en la España de los siglos XVI y XVII," *Anthropos*, 20: 28–36 (1982). The standard monographic study is José María López Piñero, *Ciencia y Técnica en la Sociedad Española de los Siglos XVI y XVII* (Barcelona: Editorial Labor, 1979), to which may be added his *La Introducción de la Ciencia Moderna en España* (Barcelona: Ediciones Ariel, 1969).

46   I have discussed this problem earlier in "Chemistry and Iatrochemistry in Early Eighteenth-Century Portugal: A Spanish Connection," in *Historia e Desenvolvimento da Ciência em Portugal: I Colóquio – até ao Século XX. Lisboa, 15 a 19 Abril de 1985*, Secretary-General of the Colloquium, Prof. Doutor António Vasconcellos Marques (2 vols, Lisboa: Academia das Ciências de Lisboa, 1986), II, 1245–1262.

47   Reprinted with an introduction by J. M. López Piñero as *El "Dialogus" [1589] del paracelsista Llorenç Coçar y la cátedra de medicamentos químicos de la Universidad de Valencia [1591]* in the *Cuadernos Valencianos de Historia de la Medicina y de la Ciencia XX* (Serie B – Textos Clásicos) (Valencia: Cátedra e Instituto de Historia de la Medicina, 1977).

48   José M. López Piñero, "La Medicina Barroco Español," *Revista de la Universidad de Madrid*, 11: 479–515; "Paracelsus and His Work in 16th and 17th Century Spain," *Clio Medica*, 8: 113–141 (1973); and "La Iatroquimica de la Segunda Mitad del Siglo XVII" in Pedro Laín Entralgo, *Historia Universal de la Medicina* (Barcelona and Madrid: Salvat Editores, 1973), IV, 279–295.

the censorship laws and personally cultivated the sciences. As a result, along with references to William Harvey, Nicholas Copernicus, and corpuscularian philosophy, we find a sudden interest in chemistry and the chemical medicine of Paracelsus, Thomas Willis (1621–1675), and Franciscus de la Boë Sylvius (1614–1672) as part of the new philosophy.

By 1679 Juan Bautista Juanini (1636–1691) was proposing iatrochemistry as the basis for a new medicine, and in 1687 Juan de Cabriada commented on the backwardness of Spanish science and called for the adoption of the "new" chemical medicine and the establishment of a Royal Academy of Science. For these authors – and others well into the eighteenth century – chemistry rather than physics was to be the key to a new science and medicine.[49] Indicative of this is the fact that Cabriada's call for an academy of science bore fruit with the founding to the Regia Sociedad de Medicina y otras Ciencias in Seville in 1697. This delayed Paracelsian influence in Spain may be illustrated in the work of Francisco Suarez de Rivera (c. 1680–1753) who published some 40 books between 1718 and 1751.[50] Having taken his M.D. in Salamanca in 1711, he later practiced in Seville and Madrid and was eventually to become physician to the Royal family. Suarez was widely read in all fields of medicine and wrote a number of books on chemical medicine. But theoretically he seems to reflect the late sixteenth century with his emphasis on the three Paracelsian principles and Natural Magic as well as his search for the alkahest, the powder of sympathy and a potable gold. Indeed, he had himself read the work of Paracelsus and recommended this author to others, stating that he, personally, had no interest in religious orthodoxy . . . he was only interested in public health.

There is no space here to discuss in any real detail the academic acceptance of chemistry nor the chemical physiology of the seventeenth-century iatro-chemists.[51] The first is for the most part a seventeenth-century development, one that I have discussed elsewhere. From the appointment of Johann Hartmann (1568–1631) at Marburg in 1609 to the end of the century when most European universities had chairs in chemistry, this was primarily a medical development. The early objections to chemically prepared medicines were overcome during the seventeenth century, and as physicians began to find

49   See also J. M. López Piñero, "La Carta Filosofica-Medico-Chymica (1687) de Juan de Cabriada, Punto de Partida de la Medicina Moderna en España," *Asclepio*, 17: 207–212 (1965).
50   There are two important monographs on this important figure: Luis S. Granjel, *Francisco Suarez de Rivera: Médico salamantino del siglo XVIII* (Salamanca: University of Salamanca, Cuadernos de Historia de la Medicina Española, Monografías IV, 1967) and José-Luis Valverde, *La Farmacia y las Ciencias Farmaceuticas en el Obra de Suarez de Rivera* (Salamanca: University of Salamanca Cuadernos de Historia de la Medicina Española, Monografías XIII, 1970). See also Debus, "Chemistry and Iatrochemistry in Early Eighteenth-Century Portugal," 1254–1257.
51   I have discussed this subject in more detail in "Chemistry and the Universities in the Seventeenth Century," *Academiae Analecta: Klasse der Wetenschappen*, 48: 13–33 (1986).

chemicals useful in their practice, they saw a need to teach chemistry to medical students. And with very few exceptions, appointments in chemistry were made in medical faculties.

But seventeenth-century medical chemists gradually extended their interests far beyond the preparation of chemical remedies. Paracelsus had discussed disease and bodily functions in terms of chemistry and chemical analogies. Van Helmont followed in a long tradition of Paracelsians when he did the same. But the primary authors of this school were to be Franciscus de la Boë Sylvius, Thomas Willis, and Raymond Vieussens (*c.* 1635–1715). Physicians, anatomists, and chemists, they laid the groundwork of a new school of iatrochemistry concerned less with chemical preparations than with the functions of the bodily organs. And since their influence coincided with the rise of the mechanical philosophy of the late seventeenth century, their work was to result in a new and perhaps more scientific debate – at least from our point of view – that was to continue well into the eighteenth century. Lester King has already shed light on the early phases of this conflict between iatrochemists and iatrophysicists,[52] but much more needs to be done in the future. The work of these medical chemists remained current much longer than is generally admitted, the *opera* of Sylvius being published as late as 1772.

Conclusion

Our discussions at this meeting are centered on "Scientific Revolutions." In regard to chemistry, we have had two revolutions to consider . . . first, the so-called Chemical Revolution, which is normally associated with Lavoisier and the last half of the eighteenth century, and second, the Scientific Revolution itself, which is most frequently centered on the astronomy and the physics of motion of the sixteenth and seventeenth centuries.

Insofar as chemistry is concerned, I hope to have shown that if we are to understand the eighteenth century, we must understand the developments of the sixteenth and seventeenth centuries. It was then that there occurred a decisive break with the ancients through the work of Paracelsus and his followers. At the same time, the earlier interest in the application of chemistry to medicine was defined, and a union of the two fields was effected that was to become permanent. It was during the course of the seventeenth century, that chemistry became an academically acceptable field of study, and with few exceptions this occurred through the Medical Faculties of European universities. I would argue then that the Chemical Revolution cannot be studied profitably only as an eighteenth-century phenomenon, that is, primarily as a reaction

---

52   This is developed particularly in Lester S. King, *The Road to Medical Enlightenment 1650–1695* (London: Macdonald; New York: American Elsevier, 1970).

against phlogiston chemistry by Lavoisier and his colleagues. Rather, we must understand at least a two-phase Chemical Revolution, the first a reaction of the Paracelsians against Aristotelian and Galenic thought coupled with the general acceptance of chemical medicine . . . and this followed by the more familiar events we associate with the eighteenth century.

As a by-product of our study, we may refer to the comparative histories of the acceptance of chemistry in different parts of Europe. In France we found a deep division between the Parisian Medical Faculty, a stronghold of the Galenic establishment, and the Court physicians who leaned more toward the newer chemical remedies and doctrines. Here medical politics were combined with deeply held religious convictions — all of this leading to a sharp con- frontation. In England these problems were less divisive, and one finds a medical establishment early accepting the new medicines. But in Spain the situation was far more intense with censorship employed to maintain religious orthodoxy. Because of its suspect nature, most of the work of Paracelsus was placed on the *Index of Prohibited Books* by the early seventeenth century, and this effectively delayed the discussion of chemical medicine until the more liberal policies of the Regency of Don Juan of Austria . . . this nearly a century after chemistry had become a subject of debate in other parts of Europe. The point I wish to make here is that if we are interested in the acceptance of scientific thought, the history of chemistry teaches us that we must frequently go beyond internalist data. Here we have an example of external factors forming an integral part of our story.

The development of chemistry in this period also warns us against a con- tinuation of the traditional division of the history of science and the history of medicine. Chemistry may not be simply classified as one of the physical sciences as is so often done today. Of course Renaissance mining treatises, alchemy, and the development of inorganic compounds for medicinal purposes may support such an interpretation, but for most authors at that time chem- istry was one of the fields of medicine, and for some surely it was the only path to true medicine. I would argue that medicine was not then artificially sepa- rated from science as historians of science and medicine frequently present their subjects today.

Finally, what are we to suggest in regard to the place of chemistry in the total spectrum of the sixteenth- and seventeen-century Scientific Revolution? If we do not insist on a positivistic interpretation of the history of science . . . and if we seek rather to present the subject in the context of Renaissance and early modern intellectual history . . . we find that Paracelsus and van Helmont were major figures, the first a contemporary of Copernicus, the second a con- temporary of Galileo, who sought to overturn the ancient authorities and to establish a new science *and* medicine through chemistry and chemical analogies. The result was a lively and often acrid debate that involved ques- tions as significant as the use of observation and experiment, mathematical abstraction and quantification, and the role of religion in the interpretation of nature and educational reform. In short, the fundamental question of what the new science should be was as important to the chemists and chemical physicians of the period as it was to the mechanists. I would say then that any interpretation of the Scientific Revolution centered exclusively on astronomy, the physics of motion and the role of mathematical abstraction can lead at best to a limited understanding of the Scientific Revolution, a subject which we all agree is essential for an understanding of the modern world.

I acknowledge with gratitude the support of The Morris Fishbein Center for the Study of the History of Science and Medicine.

7

The Newtonian Achievement

The Newtonian Revolution

I. Bernard Cohen

Originally appeared as "The Newtonian Revolution," in *Revolution in Science* by I. Bernard Cohen (Cambridge, Mass., and London: The Belknap Press of Harvard University Press, 1985): 161–75.

Editor's Introduction

The figure of Isaac Newton towers over early modern science. Recent scholarship has drawn our attention to the extraordinary range of his intellectual pursuits. For example, Newton spent more time in theological investigations than on mathematics and physics. Important research has also shown that activities which historians long dismissed as embarrassing aberrations, such as Newton's alchemy, were integral to his physics; his view of matter was strongly influenced by his alchemical thought.

But the fact remains that it is not his theology or alchemy that secured Newton's legacy, but rather his contribution to mathematics and physics. In this piece taken from a larger work on the concept of revolution in science, I. Bernard Cohen – one of the most important Newton scholars of the twentieth century and, like Hooykaas, a member of the first generation of academic historians of science – describes Newton as the apex of the Scientific Revolution and enumerates his contributions. Cohen rejects the often-repeated view that Newton synthesized the work of his predecessors. In fact Newton cleared away layers of error, not just of the ancients, but of earlier seventeenth-century figures such as Kepler, Galileo, and Descartes. His work was much more than a mere synthesis.

Just as importantly, Cohen attempts to identify what it was about Newton's method that allowed him to achieve so much; this he terms Newton's style. Many scholars have noted that the title of Newton's most important work, *Mathematical Principles of Natural Philosophy*, was unusual for the times. We saw in Westman that there was a fundamental distinction between mathematics and natural philosophy, yet here was Newton doing natural philosophy through mathematics. Cohen describes here how Newton approached natural philosophy through mathematics in an entirely novel way.

Cohen argues that Newton's style is responsible for the magnitude of his accomplishments. It may seem paradoxical to us that Newton's greatest achievement, his theory of universal gravity, was greeted with profound skepticism by many of his contemporaries. For them natural philosophy identified causes. Mechanical philosophers such as the followers of Descartes insisted that true natural philosophy must provide a mechanical account of natural phenomena. They claimed that Newton's "force" of gravity was no different from the magicians' sympathies. Newton, however, refused to be drawn on what gravity actually was: "Hypotheses non fingo" (I do not fashion hypotheses), he famously stated. What enabled Newton to avoid getting bogged down in questions of what gravity was? Cohen suggests it was his approach to natural philosophy that began with mathematics, not with a preconceived natural philosophy. Newton began with a mathematical construct, a system of moving points devoid of matter. He then added mass to the points whose only attributes were that they attracted each other and determined that the points must move in elliptical orbit. He then compared his system to the real world and saw that his system accurately accounted for the real world. This alternation between mathematical construct and comparison with the real world was the essence of Newton's style. Rather than beginning with an attempt to explain what gravity is, Newton concluded simply that it must exist as this assumption predicted the real motions of the planets so accurately.

The Newtonian Revolution

I. Bernard Cohen

The Newtonian revolution differs from those other revolutions (actual or alleged) in science and in mathematics which we have been considering in that Newton was said in his own lifetime to have created a revolution. He was recognized by his contemporaries for the revolution of the calculus and for a revolution in the science of mechanics created by his *Philosophiae Naturalis Principia Mathematica*. From a historical vantage point, Newton was an extraordinary figure because he made so many fundamental contributions to different fields: pure and applied mathematics; optics and the theory of light and colors; design of scientific instruments; codification of dynamics and formulation of the basic concepts of this subject; invention of the primary concept of physical science (mass); invention of the concept and law of universal gravity and its elaboration into a new system of the universe gravity; invention of the gravitational theory of tides; and formulation of the new methodology of science. He also worked on heat, the chemistry and theory of matter, alchemy, chronology, interpretation of Scripture, and other topics. The range of his intellectual career never ceases to astonish.

The Newtonian revolution in mathematics had two aspects: the invention of the calculus (an honor he shares with Leibniz) and the application of mathematics to physics and astronomy. It was the latter which produced the Newtonian revolution in science (as opposed to a revolution in mathematics). Of course, Newton had great predecessors in the art of developing natural philosophy by mathematical principles: Stevin, Galileo, Kepler, Wallis, Hooke, Huygens. In this sense the Newtonian revolution in science was the culmination of a multiauthored effort, going back to the beginning of the Scientific Revolution, rather than the creation by Newton of something wholly new. Yet the simplest comparison of Newton's *Principia* with Kepler's *Astronomia Nova*, Galileo's *Two New Sciences*, Wallis's *Mechanics*, Hooke's writings on motion, or the treatment of accelerated motions in Huygens's treatise on the pendulum clock shows a difference of several orders of magnitude in depth, scope, and technique. It is because of the size of this quantum jump that Newton's *Principia* is the "epoch" (as Clairaut said in 1747) of a "revolution in physical science."

It is sometimes alleged that Newton created a synthesis, presumably putting together disparate ideas or principles of such scientists as Kepler, Galileo, or Hooke. But Newton's revolutionary science was hardly a melding or assembling of such ideas or principles, since in actual fact Newton's *Principia* declared their falsity. Surely a 'true' science cannot result from a mere amalgamation of false ideas and principles. Among such notions whose falsity is exhibited by Newton in the *Principia* are the following:

Kepler: the three planetary laws are "true" descriptions of the motion of the planets; a solar force exerted on those bodies diminishes directly as the distance and acts only in or near the plane of the ecliptic; the sun must be a huge magnet; because of its "natural inertia," a moving body will come to rest whenever the motive force ceases to act.

Descartes: the planets are carried around by a sea of aether moving in huge vortices; atoms do not (cannot) exist, and there is no vacuum or void space.

Galileo: the acceleration of bodies falling toward the earth is constant at all distances, even as far out as the moon; the moon cannot possibly have any influence on (or be the cause of) the tides in the sea.

Hooke: the centripetal inverse-square force acting on a body (with a component of inertial motion) produces orbital motion with a speed inversely proportional to the distance from the center of force: this speed law is consistent with Kepler's area law.

We may further observe that Newton also denied the existence of 'centrifugal' forces, which were basic to the development of Huygens's physics of motion. In their place Newton introduced a concept of 'centripetal' force, a name he chose because it was similar – though opposite in sense or direction – to Huygens's 'vis centrifuga'.

Comparison and contrast of Newton's *Principles of Philosophy* (the name he often used to refer to his book) and Descartes's *Principles of Philosophy* show the nature of the Newtonian revolution. For the critical reader one of the extraordinary aspects of Descartes's *Principles* is that it is devoid of mathematics, being largely devoted to philosophy and to philosophical principles of physics or natural philosophy. Only two of the four Parts deal with physics proper and the development of the cosmic system of vortices. Here Descartes does set forth the quantitative rules for impact which we have seen to be wrong in each example. Descartes included these rules as a subset of his third law of nature. But when Wallis published the true rules in the *Philosophical Transactions of the Royal Society*,[1] they bore the more restricted and more correct title of "Laws of Motion." Newton began his *Principles of Philosophy* with a set of "definitions" followed by "axioms or laws of motion," of which the first two correspond roughly to Descartes's first two laws of nature. Newton seems to have transformed the Cartesian "regulae quaedam sive leges naturae" into his own "axiomata sive leges motus." Newton's three laws of motion, the axioms to which he reduced the system of rational mechanics, were: (1) the principle

---

1   They were found independently also by Wren and Huygens (see Dugas 1955, ch. 5).

of inertia, that a body will persevere in its state of rest or of uniform motion straight forward unless acted on by an external force; (2) the relation of a force to its dynamical effect, that an impulsive (or continuous) external force produces a change (change in a unit time for a continuous force) in the momentum of a body in the direction of action of the force; (3) the equality of action and reaction.

Newton also transformed Descartes's title of *Principia Philosphiae* into *Philosophiae naturalis Principia mathematica*, thus boasting that in mathematicizing the principles he had constructed a natural philosophy rather than a general philosophy. Newton's *Principia* is not only mathematical in the development of the principles and in the proofs and applications of the propositions; it also sets forth a significant new mode of using mathematics in natural philosophy.

Newton's *Principia* is a remarkable book on many levels. It contains original results in pure mathematics (theory of limits and geometry of conic sections), it develops the primary concepts of dynamics (mass, momentum, force), it codifies the principles of dynamics (three laws of motion), and it shows the dynamical significance of Kepler's three laws of planetary motion and of Galileo's experimental conclusion that bodies with unequal weights will fall freely (at the same place on earth) with identical accelerations and speeds. It develops the laws of curved motions, the analysis of pendulums, and the nature of motions constrained to surfaces, and it shows how to deal with the motion of particles in continually varying force fields. Newton also indicates the way to analyze wave motions, and he explores the manner in which bodies move in various resisting mediums. The crown of all appears in the final book 3, in which he discloses the Newtonian system of the universe – regulated by gravity, by the action of a general force, of which one particular manifestation is the familiar terrestrial weight. Here Newton treats at length of the orbits of planets and their satellites, the motions and paths of comets, and the production of tides in the sea.

As an example of the new level of thought in the *Principia*, consider the motion of the moon with its apparent irregularities. For a millennium and a half, astronomers had dealt with the moon's motion by constructing geometric schemes without reference to cause. Now, Newton showed that the chief source of the 'lunar inequalities' was the phenomenon of perturbation, chiefly the result of the gravitational action of the sun as well as of the earth on the moon. With the publication of the *Principia* in 1687, it became possible to deal with this problem by starting from first principles or causes and then studying the effects. As a reviewer of the second edition of the *Prinicipia* observed, this was entirely a new way to deal with the problem.[2]

2   Newton, in fact, was not able to carry out this program fully, although he claimed to have done so. He was really successful only in accounting for what is called the variation and the nodal motion (see Cohen 1980, 76–77; Waff 1975; Chandler 1975). The review appeared in the Berlin *Acta Eruditorum*.

Perhaps the greatest triumph of all was the explanation that tides are caused by the gravitational pull of the sun and moon on the seas. "The ebb and flow of the sea," Newton declared (in bk. 3, prop. 24). "arise from the actions of the sun and moon." The magnitude of his achievement is shown by his prediction of the oblate shape of the earth on the basis of his analysis of precession and the nonsymmetrical pull of the moon on the earth's supposed equatorial bulge.

Some analysts would see the greatness of the *Principia* expressed in the commitment to an inertial physics; for Newton inertia is a property of mass. Newton is the first writer to make a clear distinction between mass and weight and to recognize, furthermore, that a body's mass has two separate and distinct aspects. Mass is a measure of the body's resistance to being accelerated or undergoing a change in its state of motion or of rest; this is its inertia. (Newton sometimes used the term 'force of inertia' or 'vis inertiae' – but this type of force differs from forces that are 'active' and that can produce accelerations.) But a body's mass is also a measure of the body's response to a given gravitational field. But why should there be a relation between a body's (inertial) resistance to acceleration and its (gravitational) response to a gravitational field? In classical physics there is no reason. Newton had the insight to recognize that this relation must rest on the foundation of experiment, and so he proceeded to prove by experiment this constancy between inertia and gravity. It is only in Einstein's relativity theory that there is a logical necessity for this equivalence of 'inertial' mass and 'gravitational' mass. Einstein greatly admired Newton for having had so deep an insight into this problem and for having recognized that the only Newtonian grounds for this equivalence were experimental.

The nature of the mathematics in Newton's *Principia* is often misunderstood. A superficial turning of the pages gives the impression that the mathematics used by Newton is geometry, particularly Greek geometry. The style seems to be that of Euclid or Apollonius. But a closer examination shows that Newton is developing the subject by the calculus, by stating relations geometrically in ratios and proportions and at once considering the 'limit' as a fundamental quantity vanishes (or is nascent). Hence, although Newton does not develop an algorithm of the calculus (or 'fluxions') which he then applies systematically, he does make extensive use of limiting procedures which are clearly equivalent to using the calculus or which can readily be translated into the symbolism of either the Newtonian or the Leibnizian algorithm. Recognizing this aspect of the *Principia*, the Marquis de l'Hôpital observed (as Newton proudly noted) that the mathematics of the book is almost entirely the calculus. This would be further evident to any careful reader from the development of the theory of limits in section 1 of book 1 and from the explicit theory of fluxions (the Newtonian version of the differential calculus) in section 2 of book 2. Additionally, the *Principia* was notable for other original uses of mathematics such as the extensive use of infinite series.

Newton's Style

The essence of Newton's revolutionary science, as I see it, is to be found in what I have called the 'Newtonian style'. This can be seen most easily in Newton's treatment of Kepler's laws in the *Principia*.[3] Newton begins with a purely mathematical construct or imagined system – not merely a case of nature simplified but a wholly invented system of the sort that does not exist in the real world at all. Here by 'real' world is meant only the external world as revealed by experiment and observation. In this system or construct, a single mass-point moves about a center of force. Newton shows by mathematics (bk. 1, prop. 1) that if in this construct or system a force is constantly directed from the orbiting mass point or particle to the immobile center of force, then Kepler's law of areas (his second law) will hold. He next proves the converse (prop. 3), that if the law of areas holds there must be such a centripetal or centrally directed force. Hence the existence of a centripetal force is proved to be both a necessary and sufficient condition for Kepler's law of areas. Then Newton shows that if the orbit is an ellipse, the central force must vary inversely as the square of the distance. Finally he proves that if under such a condition of force there are several orbiting mass points, which do not interact with each other – or (what comes to the same thing) if the motion of any given mass point is compared with what its motion would be at a somewhat different distance from the center – then Kepler's third or harmonic law will hold. Incidentally, we may observe that Newton has shown here for the first time the dynamical significance of each of Kepler's laws. Newton's procedure thus far constitutes a purely mathematical phase one.

In phase two, Newton compares his mental construct with the real world. At once, of course, he discovers that in the real world (for instance, in our solar system), orbiting bodies do not move about 'mathematical' centers of force but about other real bodies. The moon moves around the earth; the earth and the other planets move around the sun. Accordingly, in order to bring his mental construct or imagined system more into harmony with the real world, Newton modifies the system so that there are now two mass points. One is at the center and attracts the one which is moving in orbit, constantly drawing it away from its otherwise rectilinear inertial path. But according to the principle that to every action there must be an equal and opposite reaction (Newton's third law of motion), it follows that if the central body attracts the orbiting body, then the orbiting body must also attract the central body. Hence

---

3   This presentation of Newton's development of the law of universal gravity is an abridgment of the fuller presentation given in my *Newtonian Revolution* (1980), §§5.4–5.6. Here there may be found also an exposition of the stages of transformation of the concept of inertia leading to Newton's first law of motion.

the mental construct becomes enlarged to a system of two interacting bodies. Newton proceeds to show that under these circumstances the orbiting body does not any longer move in a simple ellipse around the central body at a focus; rather, he finds that both will move in ellipses around their common center of gravity.

This two-body system constitutes a modified phase one in which Newton once again develops mathematically the properties of his (now revised) mental construct. He next compares the modified system with the external world, a modified phase two. Of course, he finds that this system also does not conform to the real world around us. For instance, in our solar system there is not just a single planet moving around the sun but several. Accordingly, to make his mental construct conform more closely to the system of the external world, Newton moves onto yet another phase one. He introduces two or more mass points orbiting about the central mass point, not just one. It follows, again as a result of the application of Newton's third law, that each of these orbiting mass points both is attracted by the central body and attracts it. In other words, a consequence is that each orbiting mass point is both a body that can be attracted and a center of attractive force. Each of these orbiting bodies will act upon and be acted upon by every other orbiting body. The system contains bodies which act by perturbations on one another, and these perturbations produce a slight departure from Kepler's laws. Newton then proceeds to find the quantitative measure of the deviation from Kepler's laws in our solar system.

In this kind of contrapuntal alternation between mathematical constructs and comparisons with the real world, between a phase one and a phase two, Newton advances from a one-body system not only to a many-body system but also to a system of orbiting bodies which have satellites, such as the moons of the earth, Saturn, and Jupiter. Thus far he has been considering mass points rather than physical bodies, because he has not yet introduced considerations of size and shape, but eventually he shifts the level of discussion from mass points to physical bodies with significant dimensions and figures.

The progression I have described is not merely a twentieth-century after-the-fact analysis of the way Newton presents his subject in the *Principia*. It also corresponds to the documented stages of development of Newton's ideas.[4] In the autumn of 1684 Newton wrote a tract (*De Motu*) in which he presented the results of his study of Kepler's laws and other aspects of the subject. There he shows that a central force is a necessary and sufficient con-

4   I assume here, in the absence of any contradictory evidence, that in the successive versions of his preliminary tract *De Motu* and in the *Principia*, Newton was presenting his ideas and results more or less in the logical-chronological order in which he had developed them. See Cohen 1980, 248ff., 258ff.

dition for the law of areas, and that an elliptical orbit implies that the force varies as the inverse square of the distance, much as in the later *Principia*. But he has not as yet recognized that his proofs apply only to a mental construct of a one-body system and so he proudly writes: "Scholium: Therefore the major planets revolve in ellipses having a focus in the center of the sun and by radii drawn [from the planets] to the sun describe areas proportional to the times, entirely as Kepler supposed." Before long Newton realized that the planets cannot in fact move in simple Keplerian elliptical orbits. He saw that his results apply only to an artificial one-body system in which the earth is reduced to a mass point and the sun to an immobile center of force.

In December 1684 Newton completed a revised draft of *De Motu* that describes planetary motion in the context of an interactive many-body system. Unlike the earlier draft, the revised one concludes that "the planets neither move exactly in ellipses nor revolve twice in the same orbit." This conclusion led Newton to the following result:

> There are as many orbits to each planet as it has revolutions, as in the motion of the Moon, and each orbit depends on the combined motions of all the planets, not to mention the actions of all these on one another . . . To consider simultaneously the causes of so many motions and to define the motions themselves by exact laws allowing of convenient calculation exceeds, unless I am mistaken, the power of the entire human intellect.

Newton had come to perceive that the planets act gravitationally on one another. The passage cited above expresses this perception in unambiguous language: "eorum omnium actiones in se invicem" (the actions of all of them on one another). A consequence of this mutual gravitational attraction is that all three of Kepler's laws are not strictly true in the world of physics but are true only for a mathematical construct in which masses that do not interact with one another orbit either a mathematical center of force or a stationary attracting body. The distinction Newton draws between the realm of mathematics, in which Kepler's laws are truly laws, and the realm of physics, in which they are only "hypotheses" (or approximations), is one of the revolutionary features of Newtonian celestial dynamics.

In an early draft of what was to become book 3 of the *Principia*, Newton showed how considerations of the third law of motion led to the concept of a mutual force between the sun and each planet, between a planet and its satellites, and between any two planets. The same considerations lead to the revolutionary new idea that any and all bodies in the universe must "attract one another." He proudly presented this conclusion with the explanatory comment that in any pair of terrestrial bodies the magnitude of the attractive force is so small that it is unobservable. "It is possible" he wrote, "to observe these forces only in the huge bodies of the planets." Of all the planets, Jupiter and Saturn are the most massive, and so he sought orbital perturbations in their motions. With the help of John Flamsteed, Newton found that the orbital motion of Saturn is indeed perturbed when the two planets are closest together.

In book 3 of the *Principia*, which is concerned with the system of the world but is somewhat more mathematical than the earlier version, Newton treats the topic of gravitation in essentially the same way. First, in what is called the moon test, he extends the weight force, or terrestrial gravity, to the moon and demonstrates that the force varies inversely with the square of the distance. Then he identifies the same terrestrial force with the force of the sun on the planets and the force of a planet on its satellites. All these forces he now calls gravity. With the aid of the third law of motion he transforms the concept of a solar force on the planets into the concept of a mutual force between the sun and the planets. Similarly, he transforms the concept of a planetary force on the satellites into the concept of a mutual force between planets and their satellites and between satellites. The final transformation is the notion that all bodies interact gravitationally.

My analysis of the stages of Newton's thinking should not be taken as diminishing the extraordinary force of his creative genius; rather, it should make that genius plausible. The analysis shows Newton's fecund way of thinking about physics, in which mathematics is applied to the external world as it is revealed by experiment and critical observation. Because he did not assume that the construct is an exact representation the of physical universe, he was free to explore the properties and effects of a mathematical attractive force even though he found the concept of a grasping force "acting at a distance" to be abhorrent and not admissible in the realm of good physics. Next he compared the consequences of his mathematical construct with the observed principles and laws of the external world such as Kepler's law of areas and law of elliptical orbits. Where the mathematical construct fell short Newton modified it. This way of thinking, which I call the Newtonian style, is captured by the title of Newton's great work: *Mathematical Principles of Natural Philosophy*.

The law of universal gravitation explains why the planets follow Kepler's laws approximately and why they depart from the laws in the way they do. It was the law of universal gravitation which demonstrated why (in the absence of friction) all bodies fall at the same rate at any given place on the earth and why the rate varies with elevation and latitude. The law of gravitation also explains the regular and irregular motions of the moon, provides a physical basis for understanding and predicting tidal phenomena, and shows how the earth's rate of precession, which had long been observed but not explained, is the effect of the moon's pulling on the earth's equatorial bulge. Since the mathematical force of attraction works well in explaining and predicting the observed phenomena of the world, Newton decided that the force must "truly exist" even though the received philosophy to which he adhered did not and could not allow such a force to be part of a system of nature. And so he called for an inquiry into how the effects of universal gravity might arise.

Although Newton at times thought universal gravity might be caused by the impulses of a stream of aether particles bombarding an object or by variations in an all-pervading aether, he did not advance either of these notions in the *Principia* because, as he ultimately said, he would "not feign hypotheses" as physical explanations. The Newtonian style had led him to a mathematical concept of universal force, and that style led him to apply his mathematical result to the physical world even though it was not the kind of force in which he could believe.

Some of Newton's contemporaries were so troubled by the idea of an attractive force acting at a distance that they could not begin to explore its properties, and they found it difficult to accept the Newtonian physics. They could not go along with Newton when he said he had not been able to explain how gravity works but that "it is enough that gravity really exists and suffices to explain the phenomena of the heavens and the tides." Those who accepted the Newtonian style fleshed out the law of universal gravity, showed how it explains many other physical phenomena, and demanded that an explanation be sought of how such a force could be transmitted over vast distances through apparently empty space. The Newtonian style enabled Newton to study universal gravity without premature inhibitions that would have blocked his great discovery. The eighteenth-century biologist Georges Louis Leclerc de Buffon once wrote that a man's style cannot be distinguished from the man himself. In the case of Newton his greatest discovery cannot be separated from his style.

Acceptance of a Newtonian Revolution

There are numerous testimonials to the Newtonian revolution in science. The eighteenth-century historian of science Jean-Sylvain Bailly wrote that "Newton overturned or changed all ideas": his "philosophy brought about a revolution." Bailly was not content merely to state generalities concerning the Newtonian revolution in science. As he saw it, the key that in Newton's hands unlocked the celestial mysteries was mathematics: geometry. As Bailly put it: "What is supposed to make things move is what really makes things move; the demonstration was complete. Newton alone, with his mathematics [géométrie], divined the secret of nature."

With rare insight, Bailly saw that "the advantage of mathematical solutions is that they are general." The argument that if the planets move according to Kepler's laws, they must be "impelled by a force residing in the sun" depends only on mathematical or geometrical considerations and general principles of motion. No special physical properties of the sun appear in Newton's argument, which differs from Kepler's in that the latter had invoked such special qualities of the sun as its magnetic force and the orientation of its poles. Accordingly, the identical mathematical argument shows that the satellites of Jupiter and Saturn, subject to the same laws of Kepler, must be equally "impelled by forces residing in these two planets." In other words, Jupiter and Saturn are to their satellite systems what the sun is to the planetary system, the only difference being one of extent and power. And the same is true of the earth and our moon (Bailly 1785, vol. 2, bk. 12, sec. 9, pp. 486f.).

Bailly himself was willing to accept the concept and principle of a universal gravitating force, since so many phenomena were explained by its use: so many of the observed data and experiential laws could be derived by mathematics from the properties of universal gravity (sec. 4, pp. 555f.). He was aware, however, that at first many scientists (notably in France) made a distinction between the Newtonian system as mathematical and as a true natural philosophy. Thus with respect to Maupertuis, who (according to Bailly) "appears to us to have been . . . the first of our mathematicians to have used the principle of attraction," Bailly (vol. 3 ("discours premier"): 7) had to point out that "at first he considered it only in relation to its calculable effects; he accepted gravitation as a mathematician, but not as a physicist." That is, Maupertuis went along with the Newtonian mathematical system or construct (our phases one and two) but would not grant that in the system of the world (phase three) Newton was necessarily dealing with quality.

In fact, in a paper "On the Laws of Attraction" (1732), Maupertuis had been very explicit on this point. "I do not at all consider," he wrote, "whether Attraction accords with or is contrary to sound Philosophy." Rather, "Here I deal with attraction only as a mathematician [géomètre]." Maupertuis was concerned with attraction only as "a quality, whatever it may be, of which the phenomena are calculable, considering it to be uniformly distributed through all the parts of matter, acting in proportion to the mass." Maupertuis, in other words, accepts the Newtonian style and is willing, as "géomètre," to follow out the mathematical consequences of a law of gravitational attraction. Since the results accord with the phenomena observed in nature, Maupertuis then asks himself as natural philosopher whether there is such a force as a physical entity, or whether there may be some other reason why bodies act as if there were such a force. If such a force does exist, it must have a cause; and we may observe that his thought is still so embedded in the mechanical philosophy that he restricts himself to two material causes of this gravitational action: some emanation from within the attracting body or some kind of matter outside the body.

A similar acceptance of the Newtonian style is found in the writings of Clairaut. Clairaut explained that "M. Newton . . . says expressly that he is using the term *attraction* only while waiting for its cause to be discovered; and in fact it is easy to judge by the treatise on the Mathematical Principles of Natural Philosophy that its only goal is to establish attraction as a fact" (Clairaut 1749, 330).

By the end of the eighteenth century, the concept of a universal gravity had become generally accepted. In the preface of his great *Mécanique céleste* (published 1799–1825), Laplace – the second Newton of this subject – began (1829, p. xxiii):

> Towards the end of the seventeenth century, Newton published his discovery of universal gravitation. Mathematicians have, since that epoch, succeeded in reducing to this great law of nature all the known phenomena of the system of the world, and have thus given to the theories of the heavenly bodies, and to astronomical tables, an unexpected degree of precision. My object is to present a connected view of these theories, which are now scattered in a great number of works. The whole of the results of gravitation, upon the equilibrium and motions of the fluid and solid bodies, which compose the solar system, and the similar systems, existing in the immensity of space, constitute the object of *Celestial Mechanics*, or the application of the principles of mechanics to the motions and figures of the heavenly bodies. Astronomy, considered in the most general manner, is a great problem of mechanics, in which the elements of the motions are the arbitrary constant quantities. The solution of this problem depends, at the same time, upon the accuracy of the observations, and upon the perfection of the analysis.

Although Laplace was endowed with a philosophical turn of mind, as evidenced by his *Philosophical Essay on Probabilities* of 1814, he did not feel any need – a century after the *Principia* – to discuss whether or not it is reasonable for a force of attraction to extend itself through space. The second 'book' of the *Mécanique céleste*, "On the Law of Universal Gravitation and the Motions of the Centres of Gravity of the Heavenly Bodies," begins with a chapter "On the Law of Universal Gravitation, deduced from observation." We are "induced," he writes (1829, 1: 249), "to consider the centre of the sun as the focus of an attractive force, which extends infinitely in every direction, decreasing in the ratio of the square of the distance." Wholly unabashed by the use of the Newtonian word 'attraction', and no longer repelled by the philosophical overtones of this word when considered at large and outside of the Newtonian context, Laplace concludes simply and straightforwardly that "the sun, and the planets which have satellites, are endowed with an attractive force, extending infinitely, decreasing inversely as the square of the distance, and including all bodies in the sphere of their activity" (p. 255). Furthermore, "analogy leads me to infer that a similar force exists generally in all the planets and comets." He has no problem with concluding "that the gravity observed upon the earth is only a particular case of a general law extending throughout the universe" and that this "attractive force" does "not appertain exclusively to its aggregated mass" but is "common to each component particle" (p. 258). He hails the Newtonian "universal gravitation" as a "great principle of nature," that "all the particles of matter attract each other in the direct ratio of their masses, and the inverse ratio of the square of their distances" (p. 259).

The success of the theory and applications of universal gravitation, or of what – since Einstein – is called 'classical' mechanics (or Newtonian mechanics), caused this subject to become the model or ideal for all the sciences. For example, much of the mid and late nineteenth-century argument about the Darwinian revolution centered on method, often focused on the question of whether or not Darwin had adhered to or abandoned the method of Newton. Scientists in as diverse fields as palaeontology and biochemistry envisioned a day when their science would have its Newton and reach the perfection of Newton's *Principia*. Why, Georges Cuvier asked in 1812, "should not natural history also one day have its Newton?" And around 1930 Otto Warburg lamented that the Newton of chemistry (for which the need had been expressed by J. H. van't Hoff and Wilhelm Ostwald in 1887) "has not yet arrived" (see Cohen 1980, 294).

The Newtonian revolution had also a tremendous ideological component, equaled perhaps by only one other scientific revolution, the Darwinian. Isaiah Berlin (1980, 144) has summed up Newton's influence:

> The impact of Newton's ideas was immense; whether they were correctly understood or not, the entire programme of the Enlightenment, especially in France, was consciously founded on Newton's principles and methods, and derived its confidence and its vast influence from his spectacular achievements. And this, in due course, transformed – indeed, largely created – some of the central concepts and directions of modern culture in the west, moral, political, technological, historical, social – no sphere of thought or life escaped the consequences of this cultural mutation.

Newton, and his contemporary John Locke, symbolized great new ideas, comprising that "outstanding revolution in beliefs and habits of thought" (Randall 1940, 253) that marks the modern era beginning with the Enlightenment. In contemplating this effect, we today, at three centuries' remove, sometimes find it difficult to understand how unprecedented was Newton's actual achievement in producing a mathematical theory of nature. Only adjectives like 'extraordinary' or 'phenomenal' or 'amazing' can convey the awe that scientists and nonscientists felt when Halley's Newtonian prediction that a comet would appear in 1758 (long after both Halley and Newton were dead) was verified. Men and women everywhere saw a promise that all of human knowledge and the regulation of human affairs would yield to a similar rational system of deduction and mathematical inference coupled with experiment and critical observation. The eighteenth century became "preeminently the age of faith in science" (Randall 1940, 276); Newton was the symbol of successful science, the ideal for all thought – in philosophy, psychology, government, and the science of society.

The belief in a Newtonian type of "rule of nature" according to universal laws was well expressed by the eighteenth-century physiocrats. All "social facts are linked together," according to the physiocrats, "in necessary bonds eternal, by immutable, ineluctable, and inevitable laws" (Gide and Rist 1947, 2). These would be obeyed by individuals and governments "if they were once made known to them." The physiocrats not only believed that human societies are "regulated by *natural laws*," but held that there are "the same laws that govern the physical world, animal societies, and even the internal life of every organism" (p. 8). Enlightenment men and women discarded traditional concepts of human relations and the order of human society, hoping for their Newton, who – they were sure – was "just around the corner." This "Newton of social science," according to Grane Brinton (1950, 382) would produce the new "system of social science [that] men had only to follow to ensure the *real* Golden Age, the *real* Eden – the one that lies ahead, not behind." In 1748 Montesquieu published *The Spirit of the Laws*, in which he compared a well-working monarchy with "the system of the universe," in which there is "a power of gravitation" that "attracts" all bodies to "the center." As in the model of the *Principia*, Montesquieu "laid down . . . first principles" and found that the particular cases follow naturally from them.

On almost every conceivable level of thought and action in which rational principles could be applied, the Newtonian revolution had a significant impact. Even today, when Newtonian concepts of time, space, and mass, and even the Newtonian principles of gravitation, have suffered Einsteinian replacements, there are huge areas of science and of common experience in which Newtonian science still reigns supreme. These encompass all of the experience of daily life and the machines we ordinarily use (except 'nuclear' devices). The most spectacular event of our times – the exploration of space – is not an illustration of Einsteinian relativity but only a straightforward application of classical gravitational physics – the science achieved by Newton in his *Principia* and developed by two centuries of Newtonians into the great science of rational mechanics and its central core of celestial mechanics. The Newtonian revolution was not only the apex of the Scientific Revolution, it remains one of the most profound revolutions in the history of human thought.[5]

5   The literature on Newton and the Newtonian revolution is vast, encompasing the findings of a large Newton research industry. A good entry into this area is by way of R. S. Westfall's monumental biography, *Never at Rest* (1980). On the developments in astronomy and mathematical physics that led to Newtonian science and on Newton's work in these areas, see the two books by René Dugas, *History of Mechanics* (1905 [1955]), *Mechanics in the Seventeenth Century* (1958 [1954]), and Westfall's *Force in Newton's Physics: The Science of Dynamics in the Seventeenth Century* (1971). Some of the ideas in this chapter have been presented by me in greater detail in my *The Newtonian Revolution* (1980), and in two articles: "The *Principia*, universal gravitation, and the 'Newtonian style,'" in Zev Bechler, ed., *Contemporary Newtonian Research* (1982), and "Newton's Discovery of Gravity" in the March 1981 issue of *Scientific American*, 1244: 166–179.

Major documents on the development of Newton's ideas on dynamics and celestial mechanics are available, with valuable commentaries, in *Unpublished Scientific Papers of Isaac Newton* (1962), ed. A. Rupert Hall and Marie Boas Hall; John W. Herivel's *The Background to Newton's Principia* (1965); and vol. 6 of the great edition of *The Mathematical Papers of Isaac Newton* (1974), ed. D. T. Whiteside. On the history of the *Principia*, see my *Introduction to Newton's 'Principia'* (1971). Anne Miller Whitman and I have completed a new English translation of the *Principia*, scheduled for publication in the near future.

Newton's *Opticks* is available in a convenient paperback edition; a full edition, with commentary and variant readings, prepared by Henry Guerlac, is to be published in 1985. The *Lectiones Opticae* is currently being published with a translation and commentary by Alan Shapiro. *Isaac Newton's Papers and Letters on Natural Philosophy and Related Documents* (1978 [1958]), ed. I. B. Cohen and Robert E. Schofield, contains facsimile reprints (with commentaries) of Newton's articles and related letters.

On the influence of Newton in the Enlightenment and after, the best introduction is still *The Making of the Modern Mind* (1940) by John Herman Randall, Jr. Another good general source is Herbert Butterfield's *The Origins of Modern Science* (1957 [1949]). See also Henry Guerlac's *Newton on the Continent* (1981), Margaret C. Jacob's *The Newtonians and the English Revolution* (1976), and Alexandre Koyré's *Newtonian Studies* (1965). Especially useful for the Enlightenment are Grane Brinton's *Ideas and Men: The Story of Western Thought* (1950) and Peter Gay's two volumes on *The Enlightenment* (New York, 1966–1969).

8

The Scientific Revolution and the Industrial Revolution

The Cultural Origins of the First Industrial Revolution

Margaret C. Jacob

Originally appeared as "The Cultural Origins of the First Industrial Revolution," in *Scientific Culture and the Making of the Industrial West* by Margaret C. Jacob (Oxford: Oxford University Press, 1997: 99–115).

The text and notes in this chapter have been abridged. For complete footnotes the original publication should be referred to.

Editor's Introduction

Many of the most significant figures who contributed to the developments we have been calling the Scientific Revolution worked in Britain during the later seventeenth century. Around a half-century later the Industrial Revolution began, also in Britain. The material and social technologies that grew out of the Industrial Revolution were to transform the world. *Post quid* does not necessarily mean *propter quid*, but in this case even the most casual observer is forced to ask whether the connection between the two events was purely coincidental or whether the Scientific Revolution, and in particular its British manifestation, caused the Industrial Revolution.

Historians have identified numerous factors that help to explain why the Industrial Revolution occurred when and where it did: Britain's advanced commercial and financial institutions; entrepreneurial daring in both merchant and aristocratic classes; the fencing of common lands which created both a pool of labor and surplus capital; its large maritime empire which also realized a large reserve of capital; the development of the factory system. But no account can omit the technological innovations such as the spinning jenny and, most importantly of all, the steam engine, which freed humanity from the limitations of animal, wind, and water power and permitted staggering increases in production.

We who live in the twenty-first century expect science to have a technological pay-off; that is, after all, why governments and corporations fund basic scientific research. But the connections between the Scientific and Industrial Revolutions are not quite as clear as those between biomedical research in a laboratory and the production of a new drug. Recent historical research has shown that figures such as Newton and Hooke (as well as continental figures like Galileo) were interested in solving practical problems to a far greater degree than many historians were willing to acknowledge in the middle of the twentieth century. For example, developing a way to determine longitude in order to facilitate navigation and consequently international trade was the holy grail for mathematicians and astronomers for much of the seventeenth and eighteenth centuries. But it is difficult to point at major inventions in the Industrial Revolution and show that they were the direct product of the Scientific Revolution.

In this excerpt from a larger work on the influence of the Scientific Revolution, Margaret C. Jacob argues this is the wrong way to approach the question. There was a very direct connection between the Scientific and Industrial Revolution, but it did not consist of particular inventions. Rather the link was a broader, cultural one. Because of popular textbooks, wandering lecturers, and regional educational and scientific societies which disseminated an applied version of Newtonian mechanics, the new scientific learning pervaded eighteenth-century British society to a degree far beyond that found in other European nations; engineers, mechanics, religious dissenters, elite and middle-class women were all versed in a Newtonianism that "tied science to machines" by linking principles to practice. Thus engineers seeking to apply this knowledge to solve practical problems in the digging of canals, tunnels, and mines and in the construction of machinery shared a common Newtonian language with the entrepreneurs who financed such ventures.

The Cultural Origins of the First Industrial Revolution

Margaret C. Jacob

Overture

Sometimes a single life or lives within a single family manage to embody the major themes of a book. Such is the case with the Watts – uncles, fathers, wives, sons, spanning three generations in Scotland and then England from roughly 1700 to 1800. All were interested in science; all turned to independent, entrepreneurial business and then to mechanized industry. James Watt (1736–1819) became world famous because he modified and improved the simpler steam engines of the eighteenth century and made them into the most advanced technology of the age. With his modifications patented in 1775 the engines provided unprecedented power from water and coal, replacing men and horses. They could drain deep mines and fill tidal harbors. Fitted with a patented rotary device, they ran the new cotton factories, potteries, and breweries. The steam engine became both symbol and reality of industrial changes that by the 1780s in textiles such as cotton were beginning to be seen as revolutionary.

Before James Watt became world famous he was the son of a little-known Scottish merchant, James Watt of Greenock (1698–1782) and nephew to two uncles, John and Thomas. All were in one way or another mathematical practitioners and knowledgeable about instruments and machines. One uncle, John Watt of Crawfordsdyke (1687–1737), short-lived and struggling, left the outlines of a life that, along with what is known about his more famous relatives, enliven any history book. In his hand-written notebooks, inherited from someone of the previous generation and shared with his brother, Thomas, John Watt recorded the intellectual and conceptual tools he learned from the new scientists, from Copernicus right through to his brilliant contemporary, Isaac Newton (d. 1727). Watt also inscribed his debt to the intellectual ferment associated with the English Revolution and with reforming Puritanism as it made its way after 1660 into Dissent. The Watts were all Calvinists of sorts; in Scotland and England that generally meant being a Presbyterian.

The intellectual roots of the Industrial Revolution are rudimentarily there in the jottings of John Watt, obscure artisan, self-made teacher, small-time entrepreneur. We would probably never have known about him had not his nephew, James Watt, become famous and been a compulsive saver of letters, indeed of every scrap of paper. In Britain by 1720, as we shall see in subsequent chapters, there were many artisans turned educators like John and Thomas Watt. All were obscure and made their living from applied science and mathematics. They did not have an easy time of it.

John Watt's surviving business cards are dated both 1730 and 1732 and contain a short self-portrait: "a yong man come to the Cost-side that professeth to teach . . . Mathematics . . . square and cube roots, trigonometry, navigation, sailing by the arch of a great circle, doctrine of spherical triangles with the use of both globes, astronomy, dyaling, gauging of beer and wine, surveying of land, making of globes, and these things he teacheth either arithmetrically, geometrially or instrumentally." For the date it was written the English used is old-fashioned, betraying the Scottish roots of John Watt. But his artisanal learning is prodigious and he is used to explaining things by instruments for those who possess little mathematics. Like his brother, the shipping merchant in Greenock, John Watt made a business from both land and sea, and like his brother, his handwriting suggests a man who is literate, but just.[1] Making a living as a scientific lecturer was harder in 1730 than it would be in 1780, when so many more men and women saw the value of such learning. By then, however, a kit of scientific instruments would cost about 300 pounds, a sum that John Watt probably did not see in an entire year of work.[2] Before his death a few years after he printed his business cards, John Watt got into financial trouble. We do not know why. His nephew, James Watt of engine fame probably inherited his books and used the mathematical exercises and mechanical lessons when he too learned surveying and the making of globes and quadrants.

The just-literate uncle was learned in higher scientific culture, but in his way. Aside from being literate – only slightly more than half of all Scottish men and even fewer women were at the time – he had an acquaintance with the teachings of Kepler, Copernicus, Tycho Brahe, Newton, and the mechanical philosophers. "Kepler observes that ye pulse of a strong healthful man beats about 4000 strokes in an hour . . . 67 times in a minute," Watt taught, and knowing how to count a pulse beat, a navigator at sea without a clock could roughly estimate time. One manuscript exercise book that John Watt owned started up in the 1680s; it too was probably inherited from a relative of the previous generation. It gave the phases of the moon supposedly from William the Conqueror right into the reign of Charles II (d. 1685) "whom God grant long to reign over us." Then came another page with the dates of full moons from 1687 to 1690. This book had been started sometime after the English

1   For the business cards that are stuck in a manuscript volume see JWP, BPL, MS C4/B28; for the letters of James Watt to his brother in the same collection, C4/A4, letter book for 1740–41. His account books also comprise many volumes.

2   Article by Simon Schaffer in John Brewer and Roy Porter, eds, *Consumption and the World of Goods*, New York, Routledge, 1993, p. 492.

Revolution, during the Restoration of the established church and king (1660–1685).

To show the position of earth, moon, and sun, the maker of the Watt book gave both the Copernican and the Tychonic systems. Living after 1660 he was savvy enough to know that the geocentric model of Ptolemy was, as Descartes put it in the 1640s, "now commonly rejected by all philosophers."[3] While natural philosophers in the Royal Society at this time were sure enough about the Copernican system of the sun in the center of the universe, there was still some doubt among everyday scientific practitioners. So this fellow hedged his bets and learned Tycho's system, which still put the earth in the center with elliptically orbiting planets around the sun. He also understood the completely heliocentric system of Copernicus with earth and planets revolving around the sun. For the purposes of navigation, either would do. Indeed what interests us is how this teacher of seamen and navigators was up on the latest theories about the structure of the heavens. By the 1680s the Ptolemaic system, with the earth in the center and perfect circular orbits made by planets and sun around it, was simply no longer believed. The Watt brothers were better at science than they were at history. Their knowledge of Copernicus was sketchy, perhaps recorded from memory: "Copernicus a famous astronomer of Germany, who lived in the year 1500. . . ." Actually he was a Pole who published his famous work in 1543. But no mind, the details of Copernicus's "system" were accurate enough in John Watt's handbook of applied science.

The new mechanics that evolved in the seventeenth century along with the new astronomy was synthesized into English-language textbooks written generally after 1700 by the followers of Robert Boyle and Isaac Newton. This new science, as we have seen, presumed on seeing the world, everything from air to water and earth, as composed of particles possessed of weight and measurement. In addition rational mechanics as it was developing did not abandon the traditional function of the discipline; it too organized local motions and made them more usable with the assistance of levers, weights, pulleys, and rotary motion.

Somehow John Watt and his brother Thomas had learned enough of the new mechanics to make drawings of inventions intended to be used at sea to measure the distance traveled by ship. Presumably they were the inventors. Wheels of graded circumference turned one into another, powered by the weight of water against a wheel that protruded into the sea. Carefully calibrated, each wheel reduced the feet into inches traversed, as would a series of connected pendula, and the final wheel mounted on a cabin wall would show (in ten movements of a hand on a circle) that the ship had traveled 10 miles.

---

3   Daniel Garber, *Descartes' Metaphysical Physics*, Chicago, University of Chicago Press, 1992, p. 182, citing the preface to part III of his *Principles*.

One drawing bears the signature of Thomas Watt, and it was still more sophis-ticated: "The great wheel which is to turne about once every 100 part sailing, turn yet second wheel 6 times about, and this turn ye ballance wheel 6 tymes about . . . ye index wheel turning about once in 10 tymes of this which will make a 10th part of a day. . . ."[4] It was an extremely cumbersome device, easily dislodged by the rocking of a ship. It probably never made it to the patent stage.

The drawings prove that mechanical invention occurred in the family and that early in the eighteenth century the Watts could think about the weight of water and the calibration of movement proportionally. They could also think about the smallest particles of air possessing weight as a result of their motion and they did exercises to determine "the weight of smoke that is exhaled of any combustable body." In a separate notebook probably dated 1722–1723, John Watt left a treatise on mechanic principles full of axioms and definitions: "The Center of Gravity of a Body is the point thereof about which the parts remain in equilibrio . . . velocity . . . by which a body runs a given space in a given time is the ratio of the space to the time. . . ." Watt was learning his Newtonian mechanics, possibly using a French treatise by the Dutch Newtonian s'Gravesande. In the same book he went on to apply the prin-ciples to weight balancing on a lever, to wheels, gears, etc. He is also reading physico-theology.[5]

Although learned in the latest mechanical science neither Watt uncle had even a modicum of success at inventing. Although they were teachers of mechanics, navigation, and fortification, the astrological predictions they also inherited may have meant more to them. Their notebooks contain what is described as the 1681 writings of the radical astrologer, John Pordage. What the astrologer had to say may have appealed to the precariousness of their lives, both personal and as Dissenters, political. Why else would someone in the family have copied the predictions?

Pordage was no run-of-the-mill astrologer. From the 1650s onward he was a radical in both philosophy and politics who sided with the enemies of absolute monarchy and regularly predicted dire fates for kings and potentates, even for bankers and clergy: "The conjunction of ye sun and mars will have a strange effect in some countreys in Europe & some prince perhaps last from England finds its true lot . . . some moneyed men shall suffer loss, & that from ye breaking of some great banker or bankers in or about ye city of London; some clergyman may be frowned upon by his prince."[6] The authorities of

4   JWP, BPL, MS C4/B29, n.f.
5   Muirhead MSS, BPL, MIV/box 14/1. "Essai d'une Nouvelle Theorie du Choc de Corps par Gravesande 1722," appears in a margin.
6   JWP, BPL, C4/B32, dated 1682 on cover. For background see Ann Geneva, *Astrology and the Seventeenth Century Mind. William Lilly and the Language of the Stars*, New York, Manchester Uni-versity Press, 1995; and on Pordage see Christopher Hill *The World Turned Upside Down*, London, Penguin, 1972, pp. 224–26.

church and state never liked the Pordages of their world, and after 1660 outlawed the Dissenters (non-Anglican Protestants) who were especially drawn to preachings associated with radicals like Pordage.

The year 1681 was bad for Dissenters and, as far as we know and as far back as anyone of the next generation could remember, the Watts were Dissenters. Although more numerous in Presbyterian Scotland than in most places in the kingdom, they faced persecution and now the prospect of a Catholic king. In 1681 the movement led by Whigs to exclude James, duke of York and brother of Charles II, from the throne had failed utterly. Since 1660 Dissenting clergy – Presbyterian, Congregational, especially Anabaptist and Quaker – had been jailed or fined and many had migrated to the new world or to the Dutch Republic. Although granted liberty after the Revolution of 1689, people like the Watts would remain second-class citizens throughout the eighteenth century. Not surprisingly, the same Watt notebook with the predictions contains considerable information on the colony of Pennsylvania where William Penn and the Quakers had granted everyone full religious liberty. Being attracted to the subversive preachings of Pordage and having an interest in Pennsylvania bespoke a degree of religious, if not political, radicalism in the roots of this entrepreneuring family. A hundred years later it would surface again in the revolutionary decade of the 1790s when the grandnephew of John Watt, James Watt, Jr., sided with the French revolutionaries.

A full century earlier, reading the astrologer Pordage along with the Scriptures also denoted a devout Protestantism. As Pordage said in predicting by the stars: "we do not thereby pervert ye true meaning of ye Scriptures & tho we are forbid in ye holy Scriptures to be afraid or dismayed at ye signs of heaven, viz. to be possessed with such a fear as is inconsistent with our confidence in God, or as disturbs us in performing ye duties we owe as creatures to our great creator." Another searcher of the Scripture, Isaac Newton, who preferred to take his millenarian predictions directly from his own reading and nominally an Anglican, could not have agreed more.

The Watts of Newton's lifetime illustrate the way in which we must understand science in his time, like dark thread entwined in a tapestry of many colors, a whole cloth made up of religious and secular values crisscrossed with scientific learning. Once people were literate they had resources that went from the Bible to astronomical tables; once they had some capital and some commerce they could try to take shortcuts in industrial ventures by using levers, weights, and engines. We separate science from religion, science from technology, theories from practices. They did not.

John Watt left a legacy of scientific learning and disciplined striving that never deserted the Watts for a hundred years. In the course of the eighteenth century other Europeans would arrive at the same knowledge with different values and assumptions: devotion to kings or Catholic clergy, or an aristocratic dislike of business and commerce, or a good eye for commerce and no particular interest in applied mechanisms. Of all the ways science could be woven into a wearable cloth, the way the Watts spun will remain the focus of this book. Their success was not, however, in the stars despite their interest in astrology. The economics of their situation could not predict their eventual triumph, although having access to capital was clearly essential. By the mid-eighteenth century consumption and international commerce had given the British a precious commodity in the eighteenth century, surplus capital. They also had coal, iron, and cheap labor. As we are about to see, they also possessed a distinctive scientific culture that now needs to be factored into the economic setting.

The Turn to Mechanized Industry: The Setting of Engineers and Entrepreneurs

Purely economic models traditionally assume that if people have coal, capital, and cheap labor they will see it as being in their best interests to industrialize. If they need any specialized scientific or technical knowledge to do that, they will just go out and get it. Such arguments about the way human beings change, make choices, or even recognize what choices are available, presume a particular definition of the way people are. Their free will prodded by their economic interests creates the advantageous cultural setting required, or free decultured agents simply transcend any restraints that culture may impose. Rationality means always choosing what is perceived as being in one's best interests. Put somewhat crudely, offer someone the chance to make a profit – in this case to industrialize – and they will perceive progress, do anything, invent or innovate as needed, try and try again until they succeed.[7]

What is missing in the story of early industrialization to date is any convincing cultural paradigm – a set of recognizable values, experiences, and knowledge patterns possessed among key social actors – that offers insight into the formation of the industrial mentality of the late eighteenth century.[8] According to David Landes, for the West "work has barely begun on the non-rational obstacles to innovation, on the negative influence of institutional,

---

7   For a concise summary of mechanistic concepts at work see Carlo Cipolla, ed., *The Emergence of Industrial Societies*, Fontana Economic History of Europe, Hassocks, Sussex, Harvester Press, 1976, in particular the essay by Phyllis Deane.

8   For a good critique of rational choice economics that pervade the order model see in particular, David S. Landes, "Introduction: On Technology and Growth" in Patrice Higonnet, David S. Landes, and Henry Rosovsky, eds, *Favorites of Fortune. Technology, Growth and Economic Development since the Industrial Revolution*, Cambridge, Mass., Harvard University Press, 1991, pp. 9–17; in the same volume see the example of failure in the case of Ulster in the essay by Joel Mokyr, "Dear Labor, Cheap Labor, and the Industrial Revolution."

social, and psychological attitudes."[9] The economic model of human actions finds little of interest in the differences among the various scientific cultures that emerged in eighteenth-century northwestern Europe. The model points us elsewhere, solely to supplies of capital or cheap labor, to explain Britain's extraordinary leap forward in mining, transportation, and manufacturing. The role of culture – imagined as the tinted spectacles that enhance or impede individual perception and choice, or that sharpen short-range or long-range vision – has no place in traditional economic explanations. This book seeks to remedy a deficiency in our own cultural knowledge.

Showing the marked differences between the scientific cultures found in Britain in comparison to France or The Netherlands tries to recreate the different universes wherein entrepreneurs actually lived. From there the cultural model presented here suggests that mental universes played an historical role that was important. In this chapter we will concentrate almost entirely on Great Britain in the eighteenth century, on institutions and attitudes that worked in favor of innovation. Later chapters will explore the culture of science that can be seen in other western European settings. Laying emphasis on culture should never be seen as an attempt to supplant economic factors. In a sophisticated historical account cultural and economic life should be seen as they are experienced by human beings, as intrinsically woven together.

The eighteenth-century British civil engineer or mechanician, barely a professional figure, often self-educated and self-fashioned by pioneers like Jean Desaguliers, John Smeaton, and James Watt, is the key figure in the cultural side of the story discussed in this chapter. Indebted to the scientific culture established in England by 1700, such men acquired the necessary learning to do the more advanced calculations needed to move heavy objects over hilly terrain or out of deep coal mines never before tapped. British engineers and entrepreneurs who sought to build or improve canals and harbors and invent, as well as use, steam engines, had to be able to understand one another. Too much was at stake for their partnerships to fail (as was so often the case despite their best efforts). Scientific culture anchored around the Newtonian synthesis provided the practical and increasingly accessible vocabulary.

As it turned out, both engineers and entrepreneurs were well served by knowledge of applied Newtonian mechanics. After 1687 and the publication of the *Principia* mechanics, pneumatics, hydrostatics, and hydrodynamics had all been regularized and systematized by the Newtonian synthesis. Its eighteenth-century explicators, beginning with Francis Hauksbee and Jean

9   David S. Landes, "Introduction: On Technology and Growth" in Patrice Higgonet, David S. Landes, and Henry Rosovsky, eds, *Favorites of Fortune, Technology, Growth and Economic Development since the Industrial Revolution*, Cambridge, Mass., Harvard University Press, 1991, p. 9.

Desaguliers, then wrote textbooks, which by 1750 made applied mechanical knowledge available to anyone who was highly literate in English, soon in French and Dutch.

Access to the mechanical knowledge found in the textbooks was critically important, yet the depth and breadth of its European diffusion differed widely. By the 1720s mechanical knowledge was more visible in Britain (in both England and Scotland) than anywhere else in the West; by then the British had invented what Larry Stewart calls "public science."[10] On the Continent the spread of specifically Newtonian and applied scientific knowledge to the larger public was inhibited – but not stopped – by various factors. High among them was the power of the Catholic clergy at work in the various educational establishments found, for example, in France and the Austrian Netherlands (Belgium).

In mid-eighteenth-century Britain industrial entrepreneurs in partnership with engineers merged in a preexisting setting conductive to innovation. It fostered trial and error through a common mechanical language and through relatively egalitarian interaction among and between them.[11] Both the language and the setting guaranteed trial and error, and it was (and is) absolutely essential to technological development. Engineers needed to have a hands-on familiarity with the site intended for development while speculators or local improvers also had to possess a meaningful understanding of applied mechanics to communicate with them. Such an understanding was best learned through touching or watching mechanical devices from table-top models to the real thing. Installing the wrong engine could bring bankruptcy. Applied mechanics taught by lectures, textbooks, and schoolmasters, served as the lingua franca when coal mines needed to be drained or harbors dredged or canals installed, or mechanical knowledge transferred from one industry to the next. As we saw in John Watt's notebooks, eighteenth-century textbooks of applied science slid effortlessly into technology, if for nothing else than to illustrate with weights and pulleys the principles of local motion and how they related to planetary motion. Decades before we can date the onset of industrial development fueled by power technology, its rudiments lay in the Newtonian textbooks available to literate people.

10   Larry Stewart, *The Rise of Public Science. Rhetoric, Technology, and Natural Philosophy in Newtonian Britain, 1660–1750*, Cambridge, Cambridge University Press, 1992. For the teaching of applied mathematics, i.e., hydrostatics, geometry, astronomy, surveying, and gunnery in Edinburgh as early as the Restoration period and its growing popularity, see R. H. Houston, "Literacy, Education and the Culture of Print in Enlightenment Edinburgh," *History*, (October 1993): 373–92. See also Richard S. Tompson, "The English Grammar School Curriculum in the Eighteenth Century," *British Journal of Educational Studies*, 29 (1971) 32–39. [. . .]
11   For an example of the kind of trial and error to which I refer see Basil Harley, "The Society of Arts' Model Ship Trials, 1758–1763," *The Newcomen Society for the Study of the History Engineering and Technology, Transactions*, 63 (1991–92): 53–71. [. . .]

Historians once assumed that "much of [British] technical, scientific and organizational elements were international property before 1750."[12] But the evidence drawn from formal and informal educational sites from Rotterdam to Lyon suggests that the Continental diffusion of the culture of applied mechanics was much more sporadic and uneven than has been previously imagined. It some European cases the scientific element defined as a set of laws memorized or mathematically formulated was available, but the technical elements and organizational circumstances – the informal learning, the mechanical illustrations, the hands-on use of devices, the relatively egalitarian philosophical society, the cultural "packaging" of science – differed enormously.

British scientific culture further rested on relative freedom of the press, the property rights and expectations of landed and commercial people, and the vibrancy of civil society in the form of voluntary associations for self-education and improvement. In early eighteenth-century Britain these structural transformations worked for the interests of practical-minded scientists and merchants with industrial interests. Using Newtonian science taken from those parts of the *Principia* pertaining to the mechanics of local motion, the scientists created and the merchants consumed curricula and books applicable to technological innovation. In some cases engineer scientists also developed pumps and steam engines specifically intended as early as 1710 to enable "one man to do the work of a thousand" and aimed at the marketplace of entrepreneurs.[13]

In the Royal Society of London, but especially in numerous provincial scientific and philosophical societies from Spalding to Birmingham and Derbyshire, mechanical learning formed the centerpiece of discussions, demonstrations, and lectures. Into a setting of formal, but just as important informal institutions for applied, yet experimental scientific learning, came eighteenth-century entrepreneurs, would-be engineers, governmental agents, local magistrates, even skilled artisans – all faced with economic and technological choices and receptive to new knowledge systems promising new solutions. The route out of the *Principia* (1687) to the coal mines of Derbyshire or the canals of the Midlands was mapped by Newtonian explicators who made the application of mechanics as natural as the very harmony and order of Newton's grand mathematical system.[14] As we shall see in chapter 9 when we

12   The phrase comes from the otherwise excellent introduction by Patrick O'Brien and Roland Quinault, eds, *The Industrial Revolution and British Society*, Cambridge, Cambridge University Press, 1993, p. 4.
13   Quoted from Denys Papin, *Nouvelle manière pour lever l'eau par la force du feu*, Cassel/Frankfurt, 1707, pp. 3–6, by Alan Smith, " 'Engines Moved by Fire and Water'. The Contribution of Fellows of the Royal Society to the Development of Steam Power, 1675–1733," unpublished paper dated March 10, 1995, kindly communicated by J. R. Harris.
14   For a good summary of this argument as it stood in the 1970s see D. S. L. Cardwell, "Science, Technology and Industry," in G. S. Rousseau and Roy Porter, eds, *The Ferment of Knowledge*, Cambridge, Cambridge University Press, 1980, pp. 449–83, with good insight into Smeaton. Further research has enabled historians to expand on and nuance Cardwell's arguments.

examine British settings as diverse as coal mines or select Parliamentary committees investigating the plans submitted by engineers or canal companies, after 1750 technically literate laymen and civil engineers communicated through a common scientific heritage. Their cultural universe had fashioned the "mental capital" of the first Industrial Revolution.[15]

The cultural approach emphasizes not simply the intellectual component in the British setting, the books and lectures, but also its public and social nature, how and by whom it was absorbed and deployed. The British scientific societies were populated by men of land, business, and finance. They made science innovative in application, but not necessarily in original achievements. The social and cultural setting of British science after Newton helps explain the reative absence of originality by comparison with French science.[16] Taking note of the aristocratic character of French scientific institutions and examining how it reenforced their theoretical and mathematical bent [. . .] throws the British model into sharper relief.[17]

Within an applied framework the Newtonian mechanical tradition laid particular emphasis on mechanical experimentation and actual demonstration with levers, weights, pulleys, table-top replications of engines, and so on. When turned toward application the practical and investigative style was critically important for encouraging industrial development. It tied science to machines as well as to an accessible method capable of being used by technicians and engineers who eagerly embraced the discipline and style of replication and verification. They in turn brought these practices to technological problems. Such men could simply not have understood the sharp distinction made in modern times between the scientific and the technological.

A letter of 1778 from the civil engineer, John Smeaton to James Watt concerning his steam engine, illustrates the interaction of scientific method with trial and error industrial innovation and, not least, with profit. As part of his normal way of proceeding, Smeaton explains that "to make myself master of the subject, I immediately resolved to build a small engine at home, that I could

15   For a general approach to the themes presented here see Joel Mokyr, *The Lever of Riches. Technological Creativity and Economic Progress*, New York, Oxford University Press, 1990; the phrase belonges to Ian Inkster, *Science and Technology in History. An Approach to Industrial Development*, London, Macmillan, 1991, chap. 2; Jan Golinski, *Science as Public Culture. Chemistry and Enlightenment in Britain, 1760–1820*, Cambridge, Cambridge University Press, 1992. [. . .]

16   For a recent discussion of aspects of the French scene, see C. Comte and A. Dahan-Dalmedico, "Mécanique et physique: Euler, Lagrange, Cauchy," in R. Rashed, ed., *Sciences a l'époque de la révolution française. Recherches historiques*, Paris, Blanchard, 1988, pp. 329–444. Cf. Antoine Picon, *L'Invention de l'ingenieur moderne. L'Ecole des Ponts et Chaussées 1747–1851*, vol. 1. Paris, Presses d l'École nationale des Ponts et Chaussées, 1992.

17   For his argument see the important essay that summarizes the work of Terry Shinn, "Science, Tocqueville, and the State: The Organization of Knowledge in Modern France," *Social Research*, 59 (1992): 533–66; reprinted in Margaret C. Jacob, ed., *The Politics of Western Science, 1640–1990*, Atlantic Highlands, NJ, Humanities Press, 1994. Reinforcing Shinn's approach is Eda Kranakis, "Social Determinants of Engineering: A Comparative View of France and America," *Social Studies of Science*, 19 (1989): 5–70. [. . .]

easily convert it to difference shapes for Experiments. . . . I determined to prosecute my original intentions of finding out the true *Rationale*. . . . The fact is . . . I have no account upon which I can depend, of the actual performance upon a fair and well attested experiment, of anyone of your engines. . . . If you can shew me a clear experiment . . . I should think it no trouble to go to Soho [Watt's workshop] on purpose to see it."[18] If Smeaton became convinced of the value of Watt's innovation, then the engine could be built into plans or consultations for which Smeaton was being commissioned by canal or mine developers.

With these disciplined methods of verification and replication British engineers imagined themselves to be scientists or their imitators. They could move from hands-on knowledge of machines to the application of theories drawn from mechanics, hydrostatics, or pneumatics. In addition, science and mathematics occupied their leisure and informed the education of their children, and they bought books and instruments in all fields from optics to astronomy and telescopes.[19]

In some middle-class households technical knowledge was shared by both husband and wife, as the letters between James and Annie Watt illustrate.[20] He invented the separate condenser for the steam engine; she was a chemist in her own right who sought to perfect bleaching techniques and to replicate the experiments of the French chemist Berthollet who had produced chlorine.[21] Women's participation in scientific culture, given the inequality of their status throughout the West, can be turned into one important index of its spread. From the 1730s onward there was a European-wide effort led by Newtonians like the Italian, Francesco Algarotti, to find a female audience for science. British periodicals appeared specifically aimed at making science accessible to women. This may also have had something to do with their use of surplus capital. A 1775 guide to the London stock exchange said that stockbrokers developed to assist women making investments and to represent them on the exchange floor.[22] In Birmingham where the Watts lived, mechanics appeared in the curricula of girls' schools by the 1780s.[23]

---

18   JWP, BPL, Smeaton to Boulton and Watt, 5 Feb. 1778. Underlining in the original.
19   See Musson and Robinson, *Science and Industry in the First Industrial Revolution*, [1989], chap. 5.
20   For example, see the letters in Birmingham City Library, M.II/4/2/1–34; JW to AW, 7 Jan. 1787, Paris, on his privilege being confirmed; and in the letter of JW to AW, 8 Mar. 1787, "unfortunately Mr Calverts rotative gadgeon twisted broke off just within the coupling brasses of the link. . . ." For a refreshing approach to the issue of the private and the public spheres among the middle class, see Dror Wahrman, " 'Middle-Class' Domesticity Goes Public: Gender, Class, and Politics from Queen Caroline to Queen Victoria," *Journal of British Studies*, 32, no. 4 (1993): 396–432.
21   Discussed briefly in "Memoir of Gregory Watt. Son of the Great Engineer," by James Patrick Muirhead, ms in the James Watt Papers, Birmingham Public Library.
22   Thomas Mortimer, *Everyman His Own Broker: or, A Guide to Exchange-Alley*, London, 1775.
23   David Cressy, "Literacy in Context: Meaning and Measurement in Early Modern England," in John Brewer and Roy Porter, eds, *Consumption and the World of Goods*, New York, Routledge, 1993,

By the 1780s many of the girls in Birmingham must have come from families where manufacturing and machines were commonly discussed. The mental posture of such mechanists or engineers with entrepreneurial interests might best be described as a merger of theoretical science and highly skilled artisanal craft. They knew machines from having built them, or from having closely examined them, and what is important from our perspective, they knew that machines worked best when they took into account mechanical principles learned from basic theories in mechanics, hydrostatics, and dynamics. Once learned, the theories could then be laid to one side for as long as the basic skill in metal working or mathematics remained. As the great engineer William Jessop told his inquisitive employers in the Bristol Society of Merchant Venturers, "in the earlier part of my time [I] endeavored to make myself acquainted with these Principles [respecting the discharge of water over cascades], and having been once satisfied with the result, I have, as most practical men do, discharged my memory in some measure from the Theory, and contented myself with referring to certain practical rules, which have been deduced therefrom, and corrected by experience and observation."[24] One needed the principles as well as the practices. A good workman, as Matthew Boulton put it, should "have brains as well as hands." As a frustrated French teacher of physics said in the 1790s when his school was too poor to buy the machines and devices: "Here it will be impossible to supply by [mathematical] figures in the absence of machines . . . verbal descriptions are truly insufficient in the sciences where one can only instruct by continual manipulations [of the devices]." Or as another teacher in the same national system of secondary schools put it, without the machines "I am reduced to teaching only theory."[25] In another one of these same schools, where the commitment to instilling industrial application had become part of revolutionary ideology, a French translation of the 1740s British textbook of Desaguliers was being used in the late 1790s.

In exactly the 1790s, when the French were bringing their educational system in science closer to the British model, the Society of Civil Engineers was established in London. It embodied the marriage between theory and practice, which reformers and industrialists on both sides of the Channel advocated.[26] Its membership consisted of a "first class" of engineers, a "second class" of "Gentlemen . . . conversant in the Theory or Practice, of the several Branches of Science necessary to the profession of Civil Engineer," and a third class of

pp. 314–15, diagram 17.3. But Cressy doubts that there was an "industrial revolution." For the periodicals see Eliza Haywood's *The Female Spectator* of the 1740s and *The Ladies' Diary*, and see F. Algarotti, *Sir Isaac Newton's Philosophy Explained for the Use of Ladies*, London, 1739.

24   Cited in *The Cultural Meaning of the Scientific Revolution*, pp. 232–33; from Bristol Record Office, Bright MSS, 11168(3), 15 Nov. 1790.

25   Boulton and Watt MSS, BPL, Boulton to Count Wassilieff, 19 March 1806. [. . .]

26   See James Watt Papers, BPL, C4/C6 for a printed copy of its Rules and Regulations dated April 1793 with a list of members.

"various Artists, whose professions or employments are necessary and useful to . . . Civil Engineering." Into each class fell men whom we shall meet again: in the first, James Watt and William Jessop, civil engineers (and seven others); in the second, Matthew Boulton, Watt's genteel business partner, and Sir Joseph Banks, president of the Royal Society; in the third, men about whom another book should be written, a geographer, two instrument makers, land surveyors, a millwright, an engine maker, and a printer. Although different in their "classes" (both within the society and in the larger social universe), all shared a common technical language that mechanical manuals and textbooks had helped to codify and disseminate. Only in the society did the engineers come first, ahead of their genteel betters. By the 1790s they had become leaders in the newly emergent industries.

Applied mechanics also required some mathematical training, especially in basic geometry. As British evidence suggests both engineers and entrepreneurs needed it, but they also required skilled workmen at their industrial sites who in the words of Matthew Boulton, "can forge, file, turn and fit work mathematically true."[27] Men with a basic mathematical knowledge were everywhere scarce, but rarer still on the Continent where mathematical education had not permeated as deeply into the general population as it had in Britain.[28] British schools were teaching basic mathematics, for example, algebra, geometry, surveying, mechanics, and astronomy, in some cases as early as the 1720s. Arithmetical and mathematical texts doubled during the first half of the century, their numbers peaking in the 1740s.[29] When the engineer James Watt gave instructions to his son for his education he said that "geometry and algebra with the science of calculation in general are the foundation of all useful science, without a complete knowledge of them natural philosophy is but an amusement, and without them the commonest business is tiresome."[30] He also wanted him to master physics and mechanics along with bookkeeping.[31]

The mechanical and mathematical knowledge possessed by British engineers, entrepreneurs, and even by artisans such as those who belonged to the

27   Manchester College Library, Oxford, Truro MSS, MB to Wilson, 10 Feb. 1788.
28   This source remains basic: Nicolas Hans, *New Trends in Education in the Eighteenth Century*, London, Heinemann, 1951. See also AN, Paris F17 1344/1 for complaints in the 1790s on the lack of mathematical knowledge on the part of students as young as 15 and as old as 40.
29   See John Money, "Teaching in the market-place or 'Caesar absum jam forte: Pompey aderat': the retailing of knowledge in provincial England during the eighteenth century," in John Brewer and Roy Porter, eds, *Consumption and the World of Goods*, New York, Routledge, 1993, p. 338; and Diana Harding, "Mathematics and Science Education in Eighteenth-Century Northamptonshire," *History of Education*, I (1972): 139–59, showing that by 1729 mechanics was being taught to second-year students who for the most part would have been 17; by the 1730s mechanical apparatus was used in some schools.
30   James Watt Papers, Birmingham City Library, LB/1, to James Watt, Jr, 1785.
31   Ibid., LB/1, letters to James Watt, Jr, 3 March 1785 and 3 March 1785.

Society of Civil Engineers, came from courses given by traveling lecturers, from patient study of textbooks based on the *Principia*, from handbooks for practical mechanics or textbooks used in private academies intended for artisans, or from regular attendance at the proceedings of voluntary societies like the Lunar in Birmingham, the Literary and Philosophical in Manchester, even the Royal Society in London.[32] While on the lecture circuit Desaguliers alone addressed hundreds of men and women each year who attended ten-week courses generally at a cost of two guineas. This most famous Newtonian of the 1720s and 1730s, former official experimenter for the Royal Society, then finally gathered his texts together and published *A Course of Experimental Philosophy* (1744, expanded from 1734). It put the bulk of the new mechanical knowledge into two hefty and beautifully illustrated volumes. They began with calculating the distance needed to offset disparate weights balanced on a beam, went through levers weights, pulleys, pumps, and steam engines, and ended with a verbal and pictorial description of the Newtonian universe as explicated by the law of universal gravitation. The accessibility of the British to the new mechanical knowledge can be put concretely. A young artisan like John Watt possessed as early as the 1720s a good working knowledge of rudimentary mathematics and mechanics.[33] Similarly a school master in Bristol in the same period offered his young pupils "A Train of Definitions according to the Newtonian Philosophy."[34] Even Oxford and Cambridge taught Newtonian mechanics and basic mathematics to young gentlemen while the Dissenting academics were hotbeds of scientific learning throughout much of the century.[35]

Women and the Culture of Practical Science

The industrial process seen as a culturally configured series of applications dependent on knowledge and technique might be considered an entirely

32   Alan Smith, " ' Engines Moved by Fire and Water.' The Contributions of Fellows of the Royal Society to the Development of Steam Power," summary of paper in *The Newcomen Society for the Study of the History of Engineering and Technology. Transactions*, 63 (1991–92): pp. 229–30; see also Barbara Smith, ed., *Truth, Liberty, and Religion. Essays Celebration Two Hundred Years of Manchester College*, Manchester College, Oxford, 1986; in particular, see Jean Raymond and John Pickstone, "The Natural Sciences and the Learning of English Unitarians: an Exploration of the Role of Manchester College," pp. 127–64. One such academy at Spitalfields is currently being studied by Larry Stewart.
33   Birmingham Public Library, UK, Watt MSS, MIV/14/1, a notebook entitled "Mechanic Principles" in the hand of John Watt.
34   Bristol Record Office, White MSS, no.08158, 73–81ff. It is worth noting that visting French engineers in 1789–90 who observed carpenters and rope makers believed them to work better by virtue of their education and "national character." [. . .]
35   For the most radical of these and their curriculum, which in science differed not at all from the others, see Ruth Watts, "Revolution and Reaction: 'Unitarian' academies, 1780–1800," *History of Education*, 20 (1991): 307–23.

male or masculine venture. The cultural history of the first Industrial Revolution should not, however, be gendered so exclusively. We should not miss the attitudes and values that by 1800 women were beginning to bring to scientific learning. These are hard to get at because published texts germane to the education of mechanists and entrepreneurs are overwhelmingly by men. Aside from periodicals like the *Female Spectator*, the known attendance of women at lecture courses in mechanics and electricity and their subscription to underwrite textbooks, their independent role in economic life was often not visible. Even Annie Watt has largely stayed hidden from view, her private letters only now revealing how active she was in James Watt's business life.

But early in the nineteenth century women's relative silence is more easily penetrated. Margaret Bryan broke it by publishing a textbook in mechanics, *Lectures on Natural Philosophy: The Result of Many Years' Practical Experience of the Facts Elucidated* (1806). It grew out of her years as headmistress of a girls' school outside of London. Its subscribers' list of people who put up money to finance the publication was generally filled with the names of elite women of the aristocracy and also with many unmarried women whose London addresses suggest wealth. There were, however, other female subscribers about whom little is known. The book was dedicated to Princess Charlotte of Wales and the naturalist, Charles Hutton, who encouraged the project. There is more physico-theology in the text than was usual in other comparable texts by men, and its purpose was explicitly to arm women and all readers "with a perpetual talisman," which will "guard your religious and moral principles against all innovations."[36] The truths of religion and natural philosophy possess a deep affinity, or so it is argued, and the purpose of the text was to teach girls about physics as well as "to impress them with a just sense of the attributes of the Deity." But typical of the new industrial vision, Bryan's intention was to be "not merely mechanical, but really scientific" and therefore to associate "the theoretical with practical illustrations." She presents herself as "merely a reflector of the intrinsic light of superior genius and erudition" who is translating and moderating knowledge for anyone without "profound mathematical energies." Male writers and lecturers often said similar things. She confesses to being a follower of William Paley's version of natural theology. He stood in a long tradition of Newtonian clergymen that began with Samuel Clarke,[37] and to a man they used the Newtonian universe to illustrate God's providence and beneficence.

Like Haukesbee and Desaguliers nearly a century before her, Margaret Bryan begins with Newtonian definitions of matter and gravity in the process of introducing students to the history of the new science, beginning with

36   *Lectures*, "Address to my Pupils," n.p.
37   Preface.

Galileo and on to Boyle and Newton. She then moves to fire, evaporation, and steam. The engine described is by no means state of the art; it is a Savery engine. Yet the steam engine is presented as an instrument of progress: "But for this machine we could never have enjoyed the advantages of coal fuel in our time; as our forefathers had dug the pits as far as they could go." In predictable fashion immediately follow levers, weights, and pulleys with mechanics brought to its conclusion with "Of Man as a Machine," which despite its materialist sounding title, attributes the wonderful mechanism of the human body to divine artifice. From there she went on to air pumps, atmospheric pressure, pneumatics in general, hydrostatics, hydraulics, magnetism, electricity, optics, and astronomy (on which she had written another whole book); all were illustrated by experimental demonstrations. The scientific instruction is capped off by a preachy lecture on stoicism, obedience, cheerfulness, affection, and duty. Each stands in the service of politeness.

By 1800 the British mechanical vision neatly synthesized nature with a moral economy intended for young readers both male and female, of lowly or genteel origin.[38] Routinely teachers tied it to national greatness as exemplified by decades of technological advances. As historians now seek to understand the rise of British nationalism, the success of mechanical science needs to be added to the discussion. When British soldiers were captured by the French during the Napoleonic wars they were interrogated for their manufacturing and mechanical knowledge.[39] Such knowledge arose out of a century-long culture of science originally fostered by Protestant clergymen and scientists. By the 1790s footsoldiers could possess it. An education in scientific culture such as Margaret Bryan offered gave nationalistic pride to the daughters and sons of men who belonged, in whichever "class," to the Society of Civil Engineers.

The Anglo-Irish novelist and moderate feminist, Maria Edgeworth, was such a daughter. Richard Edgeworth, her father, mentor, and friend, belonged to the civil engineering society established in London as well as to the Lunar Society in Birmingham; both father and daughter revered scientific learning and utility.[40] They identified industry and applied science as the vehicles for improvement, particularly if they could be learned by their "backward"

---

38 G. Gregory, *The Economy of Nature Explained and Illustrated on the Principles of Modern Philosophy*, London, 1804, 3 vols; vol. I, p. viii. Gregory was largely self-educated.

39 See note 10.

40 See their letters to James Watt and James Watt, Jr, in James Watt Papers, Birmingham City Library, C6/1/9: January 11, 1811, R. E. to J. W.; C6/1/37 M. E. to J. W. Oct. 1, 1811; C6/2/96, R. E. to J. W., 7 August 1813; C6/10 J. W., to M. E. 21 May 1820 (she is in Paris). And see hers of Jan. 1820 to J. W., Jr, C6/10. For a somewhat heavyhanded account of Maria and Richard Edgeworth see Elizabeth Kowaleski-Wallace, *Their Fathers' Daughters. Hannah More, Maria Edgeworth and Patriarchal Complicity*, New York, Oxford University Press, 1991, pp. 95–101, 144–45.

Irish tenants led by a patrician and educated but Protestant elite. Richard Edgeworth predicted in 1813 that "steam would become the universal Lord, and that we should in time scorn post horses."[41] Although William Strutt of the Derbyshire family of industrialists told Maria Edgeworth that mechanical learning was too dirty a business for women, he said to her that this was not for want of their ability: "Ladies are excluded . . . from Mechanics and Chemistry because accurate ideas on the subject can scarcely be acquired without dirtying their persons but in other things they are competitors."[42] As a lady she would have been the first to agree. Her novels such as *Belinda* (1801) painted gallantry as a vice and utility as a virtue; her private correspondence to the Watt family shows her keen interest in construction devices and steam. Her eagerness extended to wanting to be among the first to try the new steam boat from Holyhead to Dublin, although she did take the precaution of writing to James Watt, Jr., to ask him if he thought it was safe. Perhaps a similar knowledge of mechanical processes may have led the far more radical feminist, Mary Wollstonecraft, to argue in 1792 in her famous *Vindication of the Rights of Woman* that now women lived in an age where brute force need not predominate.

The Cultural Argument Summarized

The cultural roots of industrial technology in Britain were long, deep, and early to multiply. By 1800 so pervasive was the new scientific learning that it fueled the imagination of British entrepreneurs and feminists alike. The Royal Society of London as early as the 1680s discussed the labor-saving value of machines. Yet for an inventor or entrepreneur to get a patent in Britain up to the 1740s the bias of the authorities was overwhelming toward the argument that a device would put the poor to work, not enhance profits by reducing labor costs.[43] Indeed Desaguliers's 1744 textbook in mechanics while discussing the steam engine contains the first instance when anyone, writing in any language, spelled out in print (vol. II, p. 468) the critical insight that mechanization undertaken by engineers could enhance the profit of entrepreneurs precisely by reducing labor costs. Desaguliers's understanding of entrepreneurial industrial practice was consonant with what earlier seventeenth-century English theorists of political economy such as William Petty had explained. They looked to the marketplace as the model of human freedom. But they equated free choice with the ability to sell commodities,

41 JWP, BPL, C6/2/96, 7 August 1813 to James Watt.
42 Fitzwilliam Musuem, Cambridge, Strutt MS 48–1947; letter of 1808.
43 Royal Society, London, MSS C.P. 18, item 8, 66–80ff. Cf. Christine MacLeod, *Inventing the Industrial Revolution. The English Patent System, 1660–1800*, Cambridge, Cambridge University Press, 1988, pp. 159–60.

not with the selling of one's labor for wages, and certainly never with leisure or idleness.[44] By the 1730s an ideology of commercial development had come to be linked in the minds of some entrepreneurs with mechanical applications and Desaguliers's writings appealed directly to them. English science in the form of Newtonian mechanics directly fostered industrialization. It was not simply or merely its handmaiden as an older historical literature once claimed.

In eighteenth-century Britain the behavior and power of the landed, propertied, mercantile, and manufacturing was understood as the natural condition of all humankind. As Paul Langford puts it, "In a society dominated by property nothing could be more inimical to prevailing values than distinctions unconnected with property."[45] By 1700 within scientific circles an ideology with Baconian roots, aimed at the propertied and mercantile, had been developed. It was distinctively favorable to industrial and entrepreneurial activity. A partnership between wealth of any kind and applied science had been forged. The economic and technological wherewithal needed to make the ideology work most effectively would, however, take many decades to emerge.

When teaching about mechanics and experimentalism, the scientific lecturers of the eighteenth century reenforced the entrepreneurial interests of the middling (often higher) men and women in their audience. They officiated at the earliest marriages of convenience formed between engineers and entrepreneurs. Desaguliers interspersed mechanical practices with a discussion of the profit to be made from doing them correctly. But the mechanician had to be mathematically and naturally philosophically learned: "The contriver was a curious practical Mechanick, but no mathematician nor philosopher; otherwise he would have been able to have calculated the Power of the river." Had the trained engineer correctly calculated the volume and hence weight of the water, Desaguliers concluded, the management of power would directly have reduced costs and increased profits.[46]

Although deeply identified with propertied interests, engineers and industrial entrepreneurs had to have different skills from those of the traditionally landed or mercantile. And predictably, as Stanley Chapman has shown, eighteenth-century merchants and industrial manufacturers were not by and large the same people. British industrial entrepreneurs either had to possess

44   Richard Biernacki, *The Fabrication of Labor in Germany and Britain, 1640–1914*, Berkeley, University of California Press, 1995, pp. 222–23; cf. Richard Olson, *The Emergence of the Social Sciences, 1642–1972*, New York, Twayne, 1993, chap. 5.

45   P. Langfor, *Public Life . . .* , p. 71. And for how science played into the seventeenth-century interests of the propertied, see James R. Jacob, "The Political Economy of Science in Seventeenth-Century England," in Margaret C. Jacob, ed., *The Politics of Western Science, 1640–1990*, Atlantic Highlands, NJ, Humanities Press, 1994, pp. 19–46.

46   *A Course of Experimental Philosophy*. London, 1744, vol. II, pp. 530–31. [. . .]

technical skill or they had to be able to hire and converse with people who did.[47] They needed to assimilate applied scientific knowledge along with business skills and the Protestant values of disciplined labor and probity. As we shall see with the Watts, enlightened notions of progress and improvement also played a distinctive role in the value systems of late eighteenth-century entrepreneurs. Improvement became the watchword of the age. Its achievement relied on the coercive power of Parliament to guarantee patents or promote turnpikes and canals. In practice that meant having members of the two Houses who could understand what the engineers and entrepreneurs were trying to do.

Mechanical knowledge came to businessmen, as well as MPs, from a variety of channels. It was taught by scientific lecturers, by schoolmasters, and by self-help textbooks. It could be found even at Cambridge and Oxford. Mechanical knowledge became a centerpiece in the curriculum of the Dissenting academics that also laid great emphasis on ideologies of personal freedom, progress, property, and representative government, and on the writings of John Locke and Adam Smith.[48] A similar optimism and emphasis on "the improvement of our *Manufactures*, by the improvement of those *Arts*, on which they depend . . . chemistry and mechanism" was routinely discussed throughout the informal network of voluntary associations commonplace in towns and cities by the second half of the eighteenth century.[49]

Under the ideological umbrella pervasive in the philosophical societies emerged a new social space. The public culture of British science created, perhaps also required, a distinctive social ambiance among engineers and their employers. Collecting, experimenting at philosophical gatherings, as well as reading and discussions of literature, even the habits of sermon and lecture attending, gave engineers and entrepreneurs a common discipline and vocabulary. In this relatively egalitarian setting the civil – as distinct from the military – engineer achieved a newfound identify. He acquired skills of direct interest to men with capital to invest or commodities to move or manufacture more quickly and more expeditiously. At the same time the entrepreneur became remarkably literate in matters technical, applied, and occasionally theoretical. In explaining to a Russian count how to turn his visiting son into a manufacturer, Matthew Boulton wrote: "I also hope he will attend a course

47   Stanley Chapman, *Merchant Enterprise in Britain. From the Industrial Revolution to World War I*, Cambridge, Cambridge University Press, 1992, pp. 58–68.
48   For Parliament and improvement see P. Langford, *Public Life . . .*, pp. 139–43. See Manchester College, Oxford, exam papers, 1823, for political philosophy among Dissenters. Dissenters could not, however, sit in Parliament.
49   The quotation is from a report and address given by Thomas Barnes, D.D. "On the Affinity subsisting and extending Manufactures, by encouraging those Arts on which Manufactures principally depend," *Memoirs of the Literary and Philosophical Society of Manchester*, vol. 1, Warrington, 1785, pp. 72 et. seq.

of Experimental and Philosophical Lectures . . . when he has attained some knowledge in these sciences, I beg he will allow me the pleasure of showing him the application of some branches of them to the Manufactures and useful Arts and not return from Soho without seeing its manufactory."[50]

Brought together by a shared technical vocabulary of Newtonian origin, engineers, and entrepreneurs – like Boulton and Watt – negotiated, in some instances battled their way through the mechanization of workshops or the improvement of canals, mines, and harbors. Their mutual scientific literacy was also the source of much grief. British engineers frequently complained of the interference they encountered at an industrial site as entrepreneurs or investors proceeded to tell them how to go about their mechanical business. John Smeaton was particularly eloquent about his frustration: "The parties interfering suppose themselves competent to become Chief Engineers."[51] But Smeaton's frustration provides a critically important piece of information. By the mid-eighteenth century British entrepreneurs and speculators knew enough mechanics to think that they could stand on the river bank or at the mine shaft and tell the engineers how to do their job. For our purposes it is sufficient to know that by 1750 British engineers and entrepreneurs could talk the same mechanical talk. They could objectify the physical world, see its operations mechanically and factor their common interests and values into their partnerships.[52] What they said and did changed the Western world forever.

---

50   Boulton and Watt MSS, BPL, Russian Mint/2 L. Copy MB to Count Woronzow, 11 August 1799. Soho was their Birmingham factory.
51   Quoted in William Chapman, *Address to the Subscribers to the Canal from Carlisle to Fisher's Cross*, Newcastle, 1823, pp. 2–3,7.
52   See L. Mulligan, "Self-Scruting and the Study of Nature . . . ," *Journal of British Studies*, 35 (1996): 311–42.

9

A Dissenting View

De-centring the "Big Picture": *The Origins of Modern Science* and the Modern Origins of Science

Andrew Cunningham and Perry Williams

Originally appeared as "De-centring the 'big picture': *The Origins of Modern Science* and the modern origins of science (*BJHS*) 26: 1993): 407–32.
  The notes in this chapter have been abridged. For complete footnotes the original publication should be referred to.

Editor's Introduction

Both Hooykaas and Cohen affirm whole-heartedly the older narrative of the Scientific Revolution in the sixteenth and seventeenth centuries. Other essays that we have read have challenged various aspects of that narrative, some by making it more inclusive of disciplines beyond the mathematical and mechanical, some by using alternative definitions of revolution, all by emphasizing the need to examine the cultural and social context of science. Yet they all still see value in the concept of the Scientific Revolution, provided it is modified in the ways they suggest. But Andrew Cunningham and Perry Williams make a much more radical claim: such tweaking cannot save a concept that is flawed in such fundamental ways that it must be abandoned entirely.

The problem in their view is that the whole notion of the Scientific Revolution arose in the mid-twentieth century as the product of a particular view of science that flourished at that time. Since we no longer hold the same view of science, narratives that both adopted that concept of science and sought to explain its origins can no longer stand. After all, it is difficult to recount the origins of something that one does not believe exists. Fundamental to their critique is the notion that there is no transcendent entity called "science" that has a constant essence across time and space. Calling ancient Greek philosophy "science," for example, is the same as calling a trireme a steamship.

Cunningham and Williams give us much food for thought and debate. They identify several methodological principles that we would do well to remember in our study of history. In fact, virtually all the historians we have read would share them. But are they correct in saying that these principles force us to ditch the notion of early modern Europe's Scientific Revolution completely? They may well be correct in arguing that the origins of MODERN science lie in the age of revolutions, but the authors we have been reading do not insist that early modern science was identical to modern science. Debus, for example, argued that chemistry went through several revolutions, one in the sixteenth and seventeenth centuries, another in the late eighteenth. Cunningham and Williams rightly insist that we should pay attention to actors' categories, but philosophers and mathematicians themselves thought they were doing something radically new in the seventeenth century. And even if the way in which people were examining nature in 1700 was not the way they were doing it in 1900 or 2000, it was nevertheless very, even if not entirely, different from the way they were doing it in 1500. And what if people studied nature for different reasons from ours – does that mean it was a completely different enterprise? You and I can create sculptures for very different purposes; does that mean we are not doing sculpture? In short this will be a provocative piece for classroom discussion.

De-centring the "Big Picture": *The Origins of Modern Science* and the Modern Origins of Science

Andrew Cunningham and Perry Williams

> What had happened to him was that the ways in which it could be said had become more interesting than the idea that it could not.
>
> A. S. Byatt, *Possession*

Like it or not, a big picture of the history of science is something which we cannot avoid. Big pictures are, of course, thoroughly out of fashion at the moment; those committed to specialist research find them simplistic and insufficiently complex and nuance, while postmodernists regard them as simply impossible. But however specialist we may be in our research, however scornful of the immaturity of grand narratives, it is not so easy to escape from dependence – acknowledged or not – on a big picture. When we define our research as part of the history of science, we implicitly invoke a big picture of that history to give identity and meaning to our specialism. When we teach the history of science, even if we do not present a big picture explicitly, our students already have a big picture of that history which they bring to our classes and into which they fit whatever we say, no matter how many complications and refinements and contradictions we put before them – unless we offer them an alternative big picture.

This paper is based on the principle that big pictures are both necessary and desirable: that if our subject is to provide not merely accumulated information or discourse without meaning, but vision, growth, understanding and liberation – as our students have a right to expect of us and as we have a right to demand of ourselves – then we need to think explicitly about the overall picture of the history of science which we present and within which we work. On this principle, the problem which we now face is not the existence of big pictures in general but the continued existence of the particular big picture on which our discipline was founded, having been established in the early years of its professionalization and embodied in textbooks such as Herbert Butterfield's *The Origins of Modern Science*. The power of this old big picture, and the difficulty with which our discipline is moving away from it, is revealed by the fact that *The Origins of Modern Science* is still in print, in paperback, over forty years after it first appeared, and that students continue to have their first and most formative encounter with the subject

No visible

either through this book or through others that rely upon variously modified versions of the same big picture.[1]

That big picture is one which in principle covers the whole of human history in a single grand sweep; science is taken to be as old as humanity itself, so that the history of science can in principle run continuously from prehistoric megaliths and Bronze Age metallurgy to the human genome project. In practice, certain periods are selected for attention: for example, classical antiquity, the Middle Ages, and the early modern period. The 'scientific revolution' of the seventeenth century is regarded as a key event; Butterfield's title proclaimed that it represented 'the origins of modern science',[2] that is to say 'modern' as distinct from 'ancient' or 'medieval' science. It was supposed to mark the true beginning of the modern world and the abandonment of the ancient and medieval world, and for that reason Butterfield called it 'the supremely important field for the ordinary purposes of education', the one piece of the history of science which students of both the Arts and the Sciences should know. Butterfield and his followers were indeed successful in making the 'scientific revolution' 'supremely important'; today, it is almost unheard of for a History of Science course not to cover it in some way or another, and it is about the only piece of our discipline which has any currency in the general intellectual world.

This is now rather unfortunate, since over the past ten years the 'scientific revolution' concept has become increasingly difficult to sustain. In the first section of this paper, we survey the problems which have arisen with this concept since Butterfield's time. We realize that these will already be familiar to many readers. Our reason for rehearsing them here is to try to construct an account which explains why we are in our present position: an account which does not claim that Butterfield and his followers were idiots or worked with a defective historiography (as can so easily be done when attacking a former generation of historians), that is to say judging them by our lights and making them out to be failures, but one which acknowledges that they were brilliantly

1   Herbert Butterfield, *The Origins of Modern Science 1300–1800*, 2nd edn, London, 1957. The chief rival as an introductory textbook in English is probably *The Scientific Revolution 1500–1800: The Formation of The Modern Scientific Attitude*, New York, 1954, by Butterfield's pupil, A. Rupert Hall, now in its second edition as *The Revolution in Science 1500–1750*, Harlow, 1983. See also Thomas S. Kuhn, *The Copernican Revolution: Planetary Astronomy in the Development of Western Thought*, Cambridge, Mass., 1957; C. C. Gillespie, *The Edge of Objectivity: An Essay in the History of Scientific Ideas*, Princeton, 1960; E. J. Dijksterhuis, *The Mechanization of the World Picture*, Oxford, 1961; Richard S. Westfall, *The Construction of Modern Science: Mechanism and Mechanics*, New York, 1971.
2   The title was also used by Koyre for a series of lectures at Johns Hopkins University in 1951. See his *From the Closed World to the Infinite Universe*, Baltimore, 1957, p. ix. Compare also the subtitle of Hall, 1954, op. cit. (1). 'The origins of modern science' was also the title of Chapter 1 of Alfred North Whitehead's *Science and the Modern World*, New York, 1925, the printed version of his Lowell lectures given the same year; he too was referring to the seventeenth century.

successful in terms of *their own aims*. Those aims we reconstruct as being the formulation and establishment of a concept encapsulating the particular big picture of the history of science which they wanted to promote: a big picture which was itself based on certain assumptions about the nature of science which were common amongst science-supporting intellectuals in the years just before and just after the Second World War. What we are striving for here is a symmetrical account, which distances us from the legacy of a previous generation of historians, not by saying 'we're right and they're wrong', but by saying 'circumstances have changed'. In particular, what we are arguing is that the reason why the concept of the 'scientific revolution' is in trouble is, not that more or better research has been done, but that we have doubts about those beliefs about the nature of science and the big picture of the history of science which that concept encapsulated and promoted. The implication of this argument is that the concept of the 'scientific revolution' cannot be revived by tweaking, modification, or the addition of a few more 'social factors', or even a heavy dose of sociology of knowledge. And if the old big picture is going, then the 'scientific revolution' must go too. Indeed, trying to hold on to the concept may be damaging, for it will hinder us from developing a new big picture.

In so far as a new big picture is now being developed, inside and outside the profession of the History of Science, it seems to be pluralist in nature; that is to say, it is based on the principle that there are many possible ways of knowing and studying the world, and that science is just the particular way-of-knowing currently dominant in our culture. Hitherto, in the old big picture, all ways-of-knowing-the-world which seemed sensible to the historian have been appropriated to a single, unitary 'history of science'. In the new big picture, a general history of ways-of-knowing-the-world across the whole of human history would have to be the history of many different things, rather than of one single thing at different stages of development. Working within this new big picture, it becomes important to understand the origin of our own culture's dominant way-of-knowing, though this is now to be seen not as the origin of modern science, by contrast with ancient or medieval science, that is the transition from one stage of science to another, but as the origin of science itself – or, as we personally prefer to call it (for reasons to be explained), the *invention* of science. A number of historians in recent years have attempted to discuss the origin of science, in the new sense, but almost without exception they have located it in the period of the old 'scientific revolution', on the assumption that these canonical events, suitably reinterpreted, correspond to the changes they are trying to identify. In principle, of course, there is no reason why science should not have originated at that time; but we believe that this happens not to be the case, and in the second section of this paper we will argue that the period 1760–1848 is a much more convincing place to locate the invention of science.

We offer this sketch of the invention of science as a heuristic for teaching and research, and a contribution to the development of a new big picture, one which does not privilege one particular kind of knowledge. Its main significance, we think, is that it helps to demarcate more clearly the place in the big picture which our culture occupies. In the third section of this paper, we turn to the question of how we should deal with the remainder of the picture – that is to say, almost all of it. In the old big picture, science was taken to be a human universal, and so could act as a neutral framework on which to organize ways-of-knowing across the whole span of human history and human cultures. If we no longer assume that science is neutral and universal, a new big picture will require our vision to be jolted out of our culture and made aware of its contingency. We mention three forms of such 'de-centring', as we call it, the first of which is relatively close to existing practice, the second of which is more difficult to imagine but still, we think, quite feasible, and the third of which points towards a subject completely different from the History of Science as we know it now and which we ourselves can barely imagine, although some of the readers of this, starting from a more advanced baseline, may be able to do better – perhaps even eventually create it.

The Origins of Modern Science

Wise is the child that knows its father.

English proverb

It is now well established that our present concept of the 'scientific revolution' of the seventeenth century, whatever precedent it may have had in earlier writing, was forged by a number of scholars during the 1940s, chiefly Alexandre Koyré, whose work was then beginning to be taken up in the USA, and Herbert Butterfield, who promoted the concept in Britain, for example with his 1948 Cambridge lectures which were the basis for *The Origins of Modern Science.*[3] This first generation of those working professionally on the History of Science were generalists, not only from necessity, because there were too few of them to specialize, but because they had a big picture of the history of science which they were eager to communicate. That big picture was itself based on their view of what science was, for at this period the main reason for working on the history of science was to speak out on behalf of

3 Roy Porter, 'The scientific revolution: a spoke in the wheel?', in *Revolution in History* (ed. Roy Porter and Mikuláš Teich), Cambridge, 1986, 290–316, see 295. Precedents for the concept can be traced back to the eighteenth century; see John R. R. Christie, 'The development of the historiography of science', in *Companion to the History of Modern Science* (ed. R. C. Olby, G. N. Cantor, J. R. R. Christie and M. J. S. Hodge), London, 1990, 5–22, especially 7–9; also I. Bernard Cohen, *Revolution in Science*, Cambridge, Mass., 1985, 51–101.

science itself and to explain its nature and importance.[4] To understand fully the concept of the 'scientific revolution', and the problems which we are now having with it, we need to understand the view of science that through it they were trying to promote.

Amongst the various kinds of people who were interested in promoting science at that time, we can distinguish at least three ways of characterizing its nature, which they used in various combinations. The first was philosophical, defining science as a particular method of enquiry, producing knowledge in the form of general causal laws, preferably mathematical, as in the physical sciences, or which could be reduced to this form. This characterization of science was a legacy of nineteenth-century positivism, and it was very strong in the 'logical positivist' position articulated in the 1930s by the Vienna Circle; but even those philosophers who were anti-positivist in their stance, such as, in their different ways, Emile Meyerson and Karl Popper, continued to accept it. Few doubted that the scientific method existed, or disputed the centrality of issues such as laws, explanation and prediction to the business of defining philosophically what science was.[5]

A second, less academic, way of characterizing science was to see it in essentially moral terms, as the embodiment of basic values of freedom and rationality, truth and goodness, and the motor of social and material progress. Those who characterized science in this way saw it as the disinterested pursuit of truth, undeflected by the sway of emotions, and free from personal, political or economic interest. They generally believed that rational, scientific thought, if more generally distributed amongst the population, would put an end to misunderstanding, prejudice and social conflict, and even to fascism and totalitarianism; a classic expression of this belief was Robert Thouless's *Straight and Crooked Thinking*, first published in 1930, which had the double aim of setting out the principles of 'straight thinking', exemplified by science, and of exposing dishonest rhetorical tricks and so freeing people from manipulation by politicians.[6] The promoters of science also believed that practically applied science could bring an end to suffering and want; futuristic utopias from this period tended to show happy contented people wearing plastic clothing, free from drudgery in a world of automated factories and domestic robots, their only problem being finding things to do with their time.[7]

---

4   For very clear evidence of this in the American context, see Arnold Thackray, 'The pre-history of an academic discipline: the study of the history of science in the United States, 1891–1941', in *Transformation and Tradition in the Sciences: Essays in Honor of I. Bernard Cohen* (ed. Everett Mendelsohn), Cambridge, 1984, 395–420, especially 402–5.

5   Emile Meyerson, *Identity and Reality*, London, 1930; original French edition, 1908. K. R. Popper, *The Logic of Scientific Discovery*, London, 1959; original German edition, 1935.

6   Robert H. Thouless, *Straight and Crooked Thinking*, 1st edn, London, 1930. [. . .]

7   See, for example, the vision of the far distant future ('Everytown 2036') in the 1936 film *Things to Come*, loosely based on H. G. Wells's *The Shape of Things to Come: The Ultimate Revolution*, London,

A third way of characterizing science was as a universal human enterprise. By this, we mean that science was seen as the expression of an innate human curiosity, a general and universal desire to understand the world, that was a fundamental part of human nature and human thought throughout time and space. Many propagandists for science represented it in this way, deliberately challenging the older arts-based concept of humanity and humanism which prevailed in the university curriculum and in intellectual life generally;[8] thus George Sarton spoke of science as 'the *new* humanism', and Julian Huxley coined the phrase '*scientific* humanism' to represent his beliefs.[9] Others use the phrase 'science and civilization' in order to express their claim that science was central to human civilization properly understood – or according to some, that science *was* human civilization;[10] the best-known example of this phrase today is probably Joseph Needham's grand project on *Science and Civilisation in China*, which was originally conceived in the late 1930s.[11]

These were the three most usual characterizations of science amongst those seeking to promote it in the 1940s, and historians of science naturally incorporated elements of all of them as they developed the concept of 'the scientific revolution'. In the first place, the philosophical characterization of science led them to conceive 'the scientific revolution' as the seventeenth-century transformation of human thought into a form close to that of 'scientific knowledge' as defined by twentieth-century philosophers of science. Its defining events were taken to be those in which knowledge approximating to the ideal of general mathematical causal laws seemed to be achieved; since this ideal derived from the physical sciences, naturally enough the physical sciences supplied almost all the defining events – the Copernican revolution in astronomy, the Galilean revolution in mechanics, and the Newtonian synthesis, which

1933; the 1934 film *Plenty of Time for Play* (excerpted in the Open University television programme *The All-electric Home* written and presented by Gerrylynn Roberts, from the course A282 'Science, technology and everyday life 1870–1950'); the BBC television film *Time on our Hands*, first transmitted in 1963.

8   Thackray, op. cit. (4), 411.

9   George Sarton, *The History of Science and the New Humanism*, Bloomington, 1962, original edn 1930; Julian Huxley, 'Scientific humanism', in his *What Dare I Think? The Challenge of Modern Science to Human Action and Belief*, London, 1931, 149–77. C. P. Snow's *The Two Cultures and the Scientific Revolution* (Cambridge, 1959) was also an expression of this view; it is often forgotten that his reason for pointing out the cultural divide between the arts and the sciences was to complain that arts people did not sufficiently understand and respect the sciences. (Snow's 'scientific revolution', incidentally, was the change resulting from the application of science to industry, which he dated not earlier than 1920.)

10   Thackray, op. cit. (4), 401, 408. The identification of science and human thought was also common; for example, 'it was as though *science or human thought* had been held up by a barrier until this moment [i.e. until 'the scientific revolution']' (Butterfield, op. cit. (1), 7, our emphasis).

11   Cambridge, 1954–, i, 11. Jacob Bronowski's *The Ascent of Man* was commissioned by the BBC as a counterpart to a similarly epic television series on the history of art by Sir Kenneth Clark; J. Bronowski, *The Ascent of Man*, London, 1973, 13. [. . .]

were conceived in terms of mathematization and the 'mechanization of the world picture'.

At the same time, the moral characterization of science meant that the historians who developed the concept of 'the scientific revolution' attributed to these defining events liberal values such as freedom and independence of thought against superstition and Church dogma. They regarded the main theme of 'the scientific revolution' as being the elevation of experience above tradition and authority, or the rise of research and experiment as against the study of ancient texts; and this was supposedly exemplified in Copernicus's rejection of Ptolemy, Galileo's rejection of Aristotle and his challenge to the Catholic Church, and – a sole example from the 'life' sciences – Harvey's rejection of Galen. Because these historians conceived 'the scientific revolution' in terms of the advancement of free thought, they generally saw politics, religion and economic circumstances only as factors impeding its progress. But at the same time, the way in which they associated science with material advance and social harmony led them to emphasize such events as Bacon's prophecy of power over nature, and the formation of the Royal Society and the Académie Royale as instances of scientific co-operation.

Finally, the fact that these historians characterized science as a universal human enterprise meant that they conceived 'the scientific revolution' as only a revolution *within* science. Since they supposed the scientific enterprise to be a fundamental part of human nature, the history of science, for them, was in principle as long as the history of the human race itself; they counted all respectable knowledge about the natural world, wherever and whenever it was found, and however it had been produced, as some form of science. Thus despite the revolutionary nature of what they believed had happened in the seventeenth-century period, they supposed that there was still an essential continuity with the past; the ancients had engaged in essentially the same activity as the seventeenth-century heroes; they had tried to answer the same questions (such as the problem of motion), only they had not done so as well or to such an advanced level. The 'scientific revolution', despite its tremendous significance, thus consisted only of 'picking up the opposite end of the stick' or putting on 'a different kind of thinking-cap';[12] it was only a change of approach to the same, supposedly eternal, problems.

All these defining characteristics of 'the scientific revolution' now seem rather dubious. The most obvious explanation for this change is internalist (that is to say, in terms of things internal to the History of Science discipline): more recent detailed, specialized studies have undermined the general big picture of 'the scientific revolution as it was first conceived.[13] For example, with

12   Butterfield, op. cit. (1), 7, 5.
13   One problem raised by more recent research, which we are not attempting to discuss here, is that the length of 'the scientific revolution' has expanded enormously, as everyone has tried to

respect to the philosophical characterization of 'the scientific revolution', it has proved very difficult to fit developments in the 'life' sciences into the mould of mathematization and mechanism; Harvey's discovery of the circulation of the blood, once seen as the exemplary case of the physiological application of mechanical ideas (e.g. the heart as a pump) is now generally accepted to have been the result of an essentially Aristotelian investigation into the 'final cause' of the heart's motion and structure, into the ways in which this organ served the purposes of the soul.[14] Even in the physical sciences, seventeenth-century mechanical philosophy has been found to have been much less close to the twentieth-century ideal than was once imagined; seventeenth-century natural philosophers have been revealed to have described the universe in terms of not only matter and motion but also spirits and powers, the whole being dependent on a metaphysics anchored in a precise and complex theology.[15]

The moral characterization of 'the scientific revolution' too has been weakened by more recent research, as new historiographies laying emphasis on the role of 'context' or 'external factors' have made it more difficult to maintain the prime role of free, independent thought. For example, the work of Robert K. Merton started a new tradition of seeing the cardinal intellectual changes of 'the scientific revolution', in England at least, as a consequence of Puritanism (or, more generally, of Reformation and Counter-Reformation theologico-politics), and hence as a consequence of the development of capitalism.[16] At the same time, studies of the rise of the mechanical arts have revealed that practical technology played an important role in those intellectual changes,

climb on the bandwagon. Now that it has been extended to the end of the eighteenth century in order to include Lavoisier (Butterfield wrote of 'The postponed scientific revolution in chemistry'; op. cit. (1), ch. 11), and back to the high medieval period to trace the origins of Galilean mechanics (as in the work of Alistair Crombie, following Pierre Duhem), we are faced with a scientific revolution which spanned maybe five centuries. As Roy Porter has nicely put it, compared with *Ten Days that Shook the World*, this is an extraordinarily leisurely revolution (op. cit. (4), 293).

14   See, for example, C. Webster, 'William Harvey's conception of the heart as a pump', *Bulletin of the History of Medicine* (1965), 39, 508–17; Andrew Cunningham, 'Fabricius and the "Aristotle project" in anatomical teaching and research at Padua', in *The Medical Renaissance of the Sixteenth Century* (ed. A. Wear, R. K. French and I. M. Lonie), Cambridge, 1985, 195–222; Andrew Cunningham, 'William Harvey: the discovery of the circulation of the blood', in *Man Masters Nature: 25 Centuries of Science* (ed. Roy Porter), London, 1987, 65–76.

15   A key work here was J. E. McGuire and P. M. Rattansi, 'Newton and the "pipes of Pan"', *Notes and Records of the Royal Society* (1966), 21, 168–43. See also Betty Jo Teeter Dobbs, *The Foundations of Newton's Alchemy: Or 'The Hunting of the Greene Lyon'*, Cambridge, 1975. For current thinking on this subject, see Simon Schaffer, 'Occultism and reason', in *Philosophy, its History and Historiography* (ed. A. J. Holland), Dordrecht, 1985, 117–43; and Simon Schaffer, 'Godly men and mechanical philosophers: souls and spirits in Restoration natural philosophy', *Science in Context* (1987), 1, 55–85.

16   The new interest in Merton is indicated by the reprint in 1970 of *Science, Technology and Society in Seventeenth Century England*, New York, originally published in *Osiris* in 1938; and by *Puritanism and the Rise of Modern Science; The Merton Thesis* (ed. I. Bernard Cohen), New Brunswick, 1990. [. . .]

and was not simply a consequence of them. In general, it is now widely recognized that religion, politics and economics to a large extent facilitated, instead of impeding, those changes which supposedly defined 'the scientific revolution': some would say they *produced* or *were constitutive of* them.

Finally, difficulties have arisen with the characterization of 'the scientific revolution' as a stage in an enterprise as universal as (supposedly) human nature itself. A greater respect for the categories, values and enterprises of historical actors has led to the appreciation that, say, when the ancient Greek philosophers made reference to the soul or the Divine in their writings on the natural world they were not making a poor effort at modern science but were succeeding brilliantly at ancient Greek philosophy, the whole point of which was the cultivation of the soul for this life and the next. In the same way, it has been realized that when the supposed heroes of 'the scientific revolution' such as Newton used theology, mysticism, alchemy and biblical chronology in their study of the natural world this was neither insanity[17] nor a failure to be properly 'scientific' but part of a coherent attempt to reach a deeper understanding of the Christian God by studying His creation. It is now being recognized that even the post-'scientific revolution' way of studying the natural world was very different from what we now call science – so different, indeed, that it has taken a great deal of effort to recover it; and there is an increased awareness of the very deep differences between the ways the natural world has been conceived and studied in ancient, medieval, Renaissance, early modern and modern times.

New historiography and new research on the history of science, then, provides one kind of explanation for why the original defining characteristics of 'the scientific revolution' now seem dubious.[18] But a more profound explanation, we believe, can be made in terms of things *external* to the discipline, in particular the weakening of those assumptions about the nature of science which were incorporated into the concept of 'the scientific revolution' when it was first developed. First, the philosophical characterization of science has become progressively weaker since the 1960s, when doubts began to grow about the existence of a single logically-defined scientific method. The most direct challenge came from Paul Feyerabend, whose *Against Method* argued that the scientific method as defined by Popper and others simply did not work; he claimed that what *had* worked historically, in the sense of producing the scientific knowledge which we now had, was a whole variety of methods, which could not be reduced to a single logical procedure. Thomas Kuhn's *Structure of*

17   This was what Biot claimed in his entry on Newton in the *Biographie universelle*, 2nd end, Paris, 1854, xxx, 366–404, especially 390 and 401.
18   As David C. Lindberg and Robert S. Westman put it in the introduction to their edited collection *Reappraisals of the Scientific Revolution*, Cambridge, 1990, the last twenty years have seen 'highly focused studies [which] took root and began subtly to undermine the wall on which Humpty Dumpty sat' (p. xviii).

Scientific Revolutions was already providing a framework for thinking about a multiplicity of methods, by arguing that all aspects of methodology, including theory evaluation, were relative to a particular scientific community's set of shared beliefs, values, techniques, and exemplary problems – their paradigm, for short.[19] There was also heavy criticism of the view that all scientific knowledge could be reduced to a single unified science in the form of general causal laws; some philosophers argued that this could not be achieved for the whole range of the physical sciences, let alone the historical sciences, such as geology, or the natural historical sciences, such as botany.[20]

In the next place, the moral characterization of science as an embodiment of the highest standards of intellectual probity and as the motor of social and material progress was also challenged by the radical, feminist and environmental movements of the 1960s, 1970s and 1980s. From a radical perspective, scientific expertise began to be seen as a form of power and control, mistrust having been prompted especially by scientists who used their authority to suppress opposition to the nuclear industry, and it began to be argued that no kind of knowledge is ever truly free of value-judgement or economic and political interest. Feminist analysis called into question the very desirability of objectivity and control over nature; when seen as a masculine attempt at emotional detachment from the world, these basic scientific ideals could seem positively pathological. And as the repercussions of the attempt to control nature became better known, and new terms such as pollution, acid rain and greenhouse effect entered the language, science and technology have ceased to be regarded as forces only for good; it became clear that they could create material problems which they were unable to solve.[21]

Finally, the characterization of science as the universal and trans-cultural knowledge-producing enterprise was weakened when this general disenchantment with modern science led to a search for ways of knowing the natural world that seemed not to share in its objectionable features. Whether a better way was sought through oriental religions, feminist epistemology, or

19 Paul Feyerabend, *Against Method: Outline of an Anarchistic Theory of Knowledge*, London, 1975; Thomas S. Kuhn, *The Structure of Scientific Revolutions*, 2nd edn, Chicago, 1970. For an interesting exegesis of Feyerabend's much-misunderstood philosophy, see José R. Maia Neto, 'Feyerabend's scepticism', *Studies in History and Philosophy of Science* (1991), 22, 543–55.
20 Jerry Fodor, 'Special sciences, or the disunity of science as a working hypothesis', *Synthese* (1974), 28, 77–115; Alan Garfinkel, *Forms of Explanation: Rethinking the Questions in Social Theory*, New Haven, 1981, ch. 2. [. . .]
21 For the radical challenge, see for example Jerome R. Ravetz, *Scientific Knowledge and its Social Problems*, Oxford, 1971; the *Radical Science Journal*, which started publication in January 1974. For the feminist challenge, see for example Hilary Rose, 'Hand, brain, and heart: a feminist epistemology for the natural sciences', *Signs: A Journal of Women in Culture and Society* (1983–4), 9, 73–90; Sandra Harding, *The Science Question in Feminism*, Milton Keynes, 1986. For the environmentalist challenge, see for example *The Limits to Growth* (ed. Club of Rome), London, 1972; *Only One Earth* (ed. Barbara Ward and Rene Dubos), London, 1972.

the traditional culture of native Americans, the appreciation of the worth of alternative knowledge forms has led to a growing sense that the pursuit of science is not a fundamental, universal feature of human nature or human civilization, but simply one among many actual or possible ways of knowing the world.

These changes in the assumptions about the nature of science over the last twenty-five years have led to various attempts to rewrite the history of the 'scientific revolution'. One set of attempts in the late 1970s, springing particularly from the radical, feminist and environmental critiques of science, followed the traditional characterization of the 'scientific revolution', but re-evaluated it as a bad thing rather than a good thing: the origin of modern forms of repression and exploitation, rather than progress and liberation.[22] Other attempts, freed by the dwindling authority of philosophy of science over historiography, sought to add qualifications and provisos to the 'scientific revolution' concept, first by the introduction of 'external factors' such as religion, politics and economics, and then insights derived from the methods of anthropology and sociology – most recently the sociology of knowledge.[23] But as more and more elements have been added on to the 'scientific revolution' concept in an effort to make it convincing, the problem has arisen that the concept is no longer obviously coherent – a problem now acknowledged quite generally, and poignantly expressed by David Lindberg and Robert Westman in their 1990 *Reappraisals of the Scientific Revolution.*[24]

We do not want to discuss here the last twenty years or so of (to use Lindberg and Westman's image) attempts to put Humpty Dumpty together again. Our argument here is that such attempts are doomed to failure, because the 'scientific revolution' concept was specifically created to encapsulate a particular big picture and a particular view of the nature of science which, while extremely convincing in the 1940s, now seem increasingly implausible. The

22   Carolyn Merchant, *The Death of Nature: Women, Ecology, and the Scientific Revolution*, San Francisco, 1980. Brian Easlea, *Liberation and the Aims of Science: An Essay on the Obstacles to the Building of a Beautiful World*, London, 1973; *Witch Hunting, Magic and the New Philosophy: An Introduction to Debates of the Scientific Revolution 1450–1750*, Brighton, 1980; *Science and Sexual Oppression: Patriarchy's Confrontation with Woman and Nature*, London, 1981.

23   For example, Steven Shapin and Simon Schaffer, *Leviathan and the Air-pump: Hobbes, Boyle, and the Experimental Life*, Princeton, 1985. Even more recently, there has appeared *The Scientific Revolution in National Context* (ed. Roy Porter and Mikuláš Teich), Cambridge, 1992, which unfortunately does not explore the question of what the scientific revolution was or whether it existed at all.

24   Op. cit. (18). The deliberate aim of the collection, and the conference which preceded it, was 'to offer at least a partial remedy' for the 'distressing situation' of the complete absence of 'a (general) picture fully consistent with recent developments [in scholarship]' (pp. xix–xx). 'Does any unity emerge' from the articles? they ask rhetorically, and conclude that 'the reader will have to decide' (p. xx). If two scholars of such calibre, after several years of effort, can find no unity to put forward, then there is little hope of the rest of us faring any better.

problem with the 'scientific revolution', we maintain, is not that insufficient research has been done, or that an up-to-date historiography is needed, or more 'external factors', or more sociology of knowledge, or discourse analysis, or whatever the current intellectual fashion happens to be. The problem is that historians are ceasing to believe in a single scientific method which makes all knowledge like the physical sciences, or that science is synonymous with free intellectual enquiry and material prosperity, or that science is what all humans throughout time and space have been doing as competently as they were able whenever they looked at or discussed nature; so small wonder that a concept developed specifically to instantiate these assumptions ceases to satisfy. If this position is correct, then (to use Nicholas Jardine's happy phrase) the 'scientific revolution' needs to be not so much rewritten as written off.[25]

This is not to say that those historians of science who have been studying the seventeenth century have been wasting their time; clearly something of great importance happened then' with respect to the investigation of nature. We are not going to make any statement here about what that something might have been. We will, however, put forward our recommendation that whatever-it-is should not be referred to as 'the scientific revolution'. People at the time spoke, either proudly or contemptuously, of 'the new philosophy', and perhaps this phrase would be a better basis for conceiving and naming what happened then; or if it is felt necessary to include the word 'revolution', then we might speak of 'the mathematical revolution', since a great change in the scope and use of mathematics was at least one of its components. But we will leave *The Origins of Modern Science* there, since we now want to turn to the question of what kind of big picture we might construct to replace the old one.

The Modern Origins of Science

It was like having a nightmare about a man who had got it into his head that τριήρης was the Greek for 'steamer', and when it was pointed out to him that descriptions of triremes in Greek writers were at any rate not very good descriptions of steamers, replied triumphantly, 'That is just what I say. These Greek philosophers . . .were terribly muddle-headed, and their theory of steamers is all wrong'. If you tried to explain that τριήρης does not mean steamer at all but something different, he would reply, 'Then what does it mean?' and in ten minutes he would show you that you didn't know; you couldn't draw a trireme, or make a model of one, or even describe exactly how it worked. And having annihilated you, he would go on for the rest of his life translating τριήρης 'steamer'.

R. G. Collingwood, *Autobiography*

---

25   Nicholas Jardine, 'Writing off the scientific revolution' (a review of Lindberg and Westman's *Reappraisals*), *Journal of the History of Astronomy* (1991), 22, 311–18.

A new big picture must take account of the changed view of the nature of science. The old big picture was based on the conception of science in the time of Butterfield and Koyré – rooted in transcendent timeless logic and embodying absolute moral values of freedom, rationality and progress: a universal human enterprise. But a new big picture must be based on the emerging re-conception of science as historically contingent and embodying the values, aims and norms of a particular social group: one amongst a plurality of ways of knowing the world. In a new big picture, what we refer to as 'science' can no longer be used as a general defining framework; it must be seen as limited, bounded in time and space and culture.

The best way of establishing such as new big picture – both of fixing it in our own minds and of teaching it to a new generation of students – is surely to focus on those boundaries: to identify the origins of science and to explain how science came into being. To describe this project as identifying the origins of science unfortunately makes it sound like the project of Butterfield and others of investigating 'the origins of modern science', so it is necessary for us to reiterate that what they claimed to have identified was the appearance of the modern form of something transcendent: the definitive and most complete realization of something which had existed in potential throughout all human history; hence their frequent use of organic metaphors such as 'birth' and 'emergence', which implied an embryonic pre-existence and the unfolding of a pre-ordained plan. By contrast, identifying the origins of science, in the revised sense, would mean finding the first appearance, the first practice, of something which is distinct and specific to our own region of time and space, rooted in the particular circumstances of our culture.

In this section we will be proposing that the origins of science in this revised sense can be located in Western Europe in the period sometimes known as the Age of Revolutions – approximately 1760–1848.[26] This proposal may be somewhat unexpected, since, as far as we are aware, all those authors to date who have discussed the origins of science in the revised sense have assumed that these coincide with 'the origins of modern science' in the old big picture; in other words, that they are to be found in the so-called 'scientific revolution'.[27] Of course the major changes in the seventeenth century, in which one may continue to believe even while abandoning the old big picture and the 'sci-

---

26   This takes its cue from the usage of E. J. Hobsbawm, *The Age of Revolution, 1789–1848*, London, 1962.
27   For example, Merchant, op. cit. (23), and Easlea, op. cit. (23), aimed to rewrite the history of the 'scinentific revolution' as the origin of the present politically-oppressive way of knowing the world. The recent wave of anti-scientistic writings consistently ascribe to the 'scientific revolution' the origin of the scientistic outlook which they criticize; see for example Bryan Appleyard, *Under-standing the Present: Science and the Soul of Modern Man*, London, Picador, 1992, especially ch. 2, 'the birth of science'. Mary Midgley's sophisticated and accessible *Science as Salvation: A Modern Myth and its Meaning*, London, 1992, is a partial exception. [. . .]

entific revolution' with it, certainly make it, at first sight at least, a plausible location for the origins of science, in the revised sense; but the extent to which the old big picture and its underlying view of science have permeated our discipline and its key concepts, such as the 'scientific revolution', makes it necessary to treat such an automatic assumption with caution. We believe that if historians of science were to come to this question of the origins of science afresh, without reference to the lines laid down by the founders of the discipline, they would locate them elsewhere. To explain why we are locating the origins of science in the Age of Revolutions, we will begin by making explicit the historiographic principles on which we have been working – principles which are, we believe, now relatively consensual within the profession.

The first of these principles, and the one that underlies and gives force to the others, is that the basic values and norms of science (its ideology, that is) are things which need explanation, rather than things which are to be taken for granted, as being above explanation, or to be used (without question) to explain other things. To put it another way, for example, the main question we used to ask about the 'objectivity' of science used to be: 'in what does the "objectivity" of science consist?'. The new question, by contrast, might be said to be: 'how did science come to have "objectivity" ascribed to it?' This principle follows from the values and norms of science no longer being seen as defined by absolute moral criteria or timeless logic, and the pursuit of science no longer being seen as a universal characteristic of human nature; if these things do not derive from some transcendent realm, then they are susceptible to, and require, historical explanation in the usual way.

The second principle is about the relation of knowledge to society, the position being that the knowledge in any society is an integral product of that society, and embodies within it the values and social relations of that society. This assumption has been used to good effect in 'New Left' work from the 1960s, and also in the more recent approach of the sociologists of science (many of whom are not politically of the Left at all).[28]

The third principle of historical enquiry concerns actors' categories. The work of Skinner and Dunn and others[29] has led us all (at least in principle) to try to respect and seek to understand the terms and categories in which the

---

28   Robert M. Young, 'The historiographic and ideological context of the nineteenth-century debate on man's place in nature', in *Changing Perspectives in the History of Science: Essays in Honour of Joseph Needham* (ed. Mikuláš Teich and Robert M. Young), London, 1973, 344–438. Steven Shapin, 'History of science and its sociological reconstructions', *History of Science* (1982), 20, 158–211.

29   Q. R. D. Skinner, 'Meaning and understanding in the history of ideas', *History and Theory* (1969), 8, 3–53. John M. Dunn, 'The identity of the history of ideas', *Philosophy* (1968), 43, 85–104. Adrian Wilson and T. G. Ashplant, 'Whig history and present-centred history', and T. G. Ashplant and Adrian Wilson, 'Present-centred history and the problem of historical knowledge', *The Historical Journal* (1988), 31, 1–16 and 253–73 respectively.

past people we study actually thought of and performed their work – in order to appreciate its authentic meaning and identity. In the recent formulation by Martin Rudwick: 'A non-retrospective narrative of any episode in the history of science should be couched in terms that the historical actors themselves could have recognised and appreciated with only minor cultural translation [to help the modern reader understand it]'[30] – and this respecting of actors' categories should not be limited to the moment when the historian is making his or her historical exposition, but should apply also to the categories within which the historian conducts his or her researches as well.

Finally the fourth principle is to do with projects of enquiry: that it is necessary to identify the particular and specific 'projects of enquiry' in which people in the past were engaged in their investigations of nature. This is a particular form of the more general principle of 'Question and Answer' (as it was originally termed by Collingwood):[31] when we read texts from the past, we need to ask ourselves, 'to what *question* – both what immediate question, and what project of enquiry – in the life and world of the person who wrote it, was this text the *answer* for its author?'. For without knowing the project that a particular historical actor was engaged on, the results arrived at by that historical actor are meaningless to us; the answer is meaningless without the question. By assigning, consciously or unconsciously, the wrong project or the wrong question to the historical actors we are investigating, the results or answers that they arrived at are given the wrong meaning by us (usually a modern-day meaning). These four historiographic principles are, we believe, now relatively consensual within the profession; for example 'Whig' or 'present-centred' history would violate all four criteria, and few would now defend it.[32]

But as well as forbidding certain things, these principles can be of positive assistance in approaching the problem of the origins of science. For instance, taking the final two together, Principle 4, about projects of enquiry, suggests that we should direct our attention, not simply to statements about the natural world in past texts, but to the precise enterprises of which these thoughts and statements were part and which gave them their identity and meaning. Principle 3, actors' categories, suggests that we should take past people's own accounts of their enterprises, including their own names for them, very seriously indeed.

Following these two principles, we can start by noticing and taking seriously the fact that it was not until the beginning of the nineteenth century that the

30   Martin J. S. Rudwick, *The Great Devonian Controversy: The Shaping of Scientific Knowledge Among Gentlemanly Specialists*, University of Chicago Press, 1985, 14.
31   R. G. Collingwood, *An Autobiography*, first published 1939, reprinted 1978, Oxford University Press, ch. 5, 'Question and answer'. A recent development of this philosophical argument is Nicholas Jardine, *The Scenes of Inquiry: On the Reality of Questions in the Sciences*, Oxford, 1991.
32   The term 'Whig' history derives from Herbert Butterfield, *The Whig Interpretation of History*, London, 1931. [. . .]

term 'science' was used for the enterprise of investigating the natural world in the way that it is used today.[33] The word 'science', deriving as it does from the Latin word *scientia*, existed prior to the nineteenth century of course. But it was not restricted to the investigation of nature, for it was used for all disciplines which dealt in terms of theory (or for the theoretical side of all disciplines), theory which was based on firm principles, and for the knowledge generated within such disciplines. Thus most of the disciplines concerned with the natural world were 'sciences', but grammar was also a science, and so was rhetoric, and so was theology. Indeed theology was often regarded as 'the Queen of the Sciences' right up until the end of the eighteenth century, whereas today it would not qualify as a science at all under our modern meaning of the term. What this difference in the application of the term 'science' should lead us to see is that for hundreds of years before this time, when western people studied the natural world they did so not as 'science' but within other disciplines, the disciplines of either 'natural history', or 'mixed mathematics' or especially 'natural philosophy'.

Yet the meaning of the most important of these, 'natural philosophy', to the historical actors themselves who actually practised it, has scarcely been investigated by historians of science. Instead it has been treated as if it was transparent: almost without exception historians of science have simply translated it as 'science' or treated it as if it meant 'science' in the modern sense, as though the two terms were interchangeable in the past;[34] or historians have thought of the term 'natural philosophy' as meaning a sort of 'general world-view', a cosmology within which the detailed, empirical investigations of science itself were performed. It is only recently that a few historians of science have regularly begun to leave the term 'natural philosophy' as it is. This has opened up the possibility of recognizing that while natural philosophy was itself indeed an investigation of the natural world which was sometimes empirical and sometimes even experimental, yet it was nevertheless one which was radically different from 'science' in the modern sense.

For the whole point of natural philosophy was to look at nature and the world *as created by God*, and as thus capable of being understood as embodying God's powers and purposes and of being used to say something about them. This is what Newton, the most famous practitioner of natural philosophy, was saying when he commented in 1692 'Et haec de Deo: de quo utique ex

33   Sadly, there have been virtually no critical discussions of the changing meaning of the word 'science'. It is probably significant that the most accessible account we know of was *not* the work of a historian of science: Raymond Williams, *Keywords: A Vocabulary of Culture and Society*, London, 1976, s.v. 'science'. There has been one good study of the word 'scientist': Sydney Ross, ' "Scientist": the story of a word', *Annals of Science* (1962), 18, 65–86 [. . .]
34   One spectacular instance we have come across is of the title of Newton's 1687 work being rendered into English as 'Mathematical Principles of Natural *Science*'.

phaenomenis disserere, ad Philosophiam experimentalem pertinet', it is the role of experimental natural philosophy to discourse of God from the phenomena. Newton shared this view of the role and purpose of natural philosophy with all other natural philosophers, as he did his belief that the study of natural philosophy, properly conceived and pursued, was a bulwark against atheism.[35] Natural philosophy scrutinized, described, and held up to admiration the universe as the true God had created it and kept it running. To the modern ear, accustomed to the distinction between 'science' and 'religion', and to a clear-cut distinction between the 'sacred' and the 'secular', this may sound as though natural philosophy was merely an aspect of theology (and particularly that it was 'natural theology'). But this was not the case: natural philosophy was an autonomous study separate from theology and from natural theology, but whose practitioners had at the forefront of their minds, as Creator of the universe they were studying, the same God whose attributes the theologians studied from other points of view. We will probably catch the appropriate attitude of reverence toward the Creation that a natural philosopher necessarily held, if we regard natural philosophy not as a sacred study, but as a godly or pious one, which could be conducted by men both in and our of holy orders.[36] To confuse natural philosophy with science is to repeat Collingwood's nightmare about the man who had got it into his head that trireme was the Greek for 'steamer'[37] The historical study of the trireme of natural philosophy is still at an early stage, but that does not mean that we can afford to interpret the term 'natural philosophy' as meaning the 'steamer' of science.

Thus the principles of 'projects of enquiry' and 'actors' categories' suggest that a major change in nomenclature applied by the historical actors themselves might well mark a change in the identity of the discipline under which they investigated nature. On this basis, a strong candidate for the origins of science, in the revised sense, should be the period when people stopped using the term 'natural philosophy' to refer to identity of their project of investigating nature, and started, for the first time, to speak of 'science' or 'the sciences' referring *only* to the sciences of nature. And this period was around the beginning of the nineteenth century.[38]

35 See Andrew Cunningham, 'How the *Principia* got its name; or, taking natural philosophy seriously', *History of Science* (1991), 29, 377–92.

36 The difficulty of finding a term which conveys the sense of Christian belief typical of early modern people and which informed the natural philosopher's view of the discipline and its subject matter (nature as created by God), but which does *not* oppose such a position to notions of objectivity, secularity, and science, is itself indicative of the distance of the identity of natural philosophy from that of science.

37 Collingwood, op. cit. (31), 64.

38 The argument of this paragraph is put in more detail in Andrew Cunningham, 'Getting the game right: some plain words on the identity and invention of science', *Studies in History and Philosophy of Science* (1988), 19, 365–89.

Now this period is one which has already been identified as highly significant in the history of science. For one thing, it was the period when many new
disciplines for the investigation of nature were created. Through Lamarck in
France and Treviranus in Germany the discipline of 'biology' was created in
the years around 1800 as a new discipline to replace the old study of the
'animal economy' or 'animated nature', which had dealt with the nature of
those things endowed with 'soul' (*anima*, in Latin); biology now covered the
same area, but now defined in terms of 'life' (*bios*, in Greek). The 'soul', the
very thing which had united animals and plants as an area of study, was to be
dismissed from this new discipline, and even to be regarded as something
'unscientific'. Geology, and its subsidiary sciences, was another new discipline
created at this period. In the hands of Cuvier and Lyell and others, an old discipline which had been described as a '*sacred* history', and in which one had
studied the earth as created and modified by God, was replaced by a *secular*
history to which questions about God, Creation and Providence were deemed
irrelevant and inappropriate. Radical transformations took place also in the
meaning and content of other old disciplines for the investigation of nature:
there was for instance a new version of physics (as Robert Fox and Maurice
Crosland have shown for Laplace and the Society of Arceuil in France, and
Susan Faye Cannon has shown for England). Similarly, there was a new version
created of that old discipline, physiology.[39]

This period has also been identified as the one in which it first became
possible for people interested in investigating the natural world to do so as a
career. Professional organizations, too, can be found at this period. Although
Britain lagged far behind France and the German states in these matters, one
can find at this time even in England the beginning of the professional career
and the first professional organizations of science, such as the British Association for the Advancement of Science (copied from a German model) or the
reformed Royal Society: 'professional' in the sense that the gentleman amateur
was beginning to be replaced by the professional (salaried) man as the model
type of person who pursued the knowledge of nature. In producing this

39   M. J. S. Hodge, 'Lamarck's science of living bodies', *BJHS* (1971), 5, 323–52. Dorinda
Outram, *Georges Cuvier: Vocation, Science and Authority in Post-Revolutionary France*, Manchester,
1984. Timothy Lenoir, *The Strategy of Life: Teleology and Mechanics in Nineteenth Century German
Biology*, Dordrecht, 1982. Roy Porter, *The Making of Geology: Earth Science in Britain 1660–1815*,
Cambridge, 1977. Susan Faye Cannon, *Science in Culture: The Early Victorian Period*, New York,
1978, especially the chapter on 'The invention of physics'. Maurice Crosland, *The Society of
Arcueil: A View of French Science at the Time of Napoleon I*, London, 1967. Robert Fox, 'The rise and
fall of Laplacean physics', *Historical Studies in the Physical Sciences* (1974), 4, 89–136. *The Inven-
tion of Physical Science: Intersections of Mathematics, Theology and Natural Philosophy since the
Seventeenth Century* (ed. Mary Jo Nye, Joan Richards and Roger Stuewer), Dordrecht, 1992. John
Lesch, *Science and Medicine in France: The Emergence of Experimental Physiology, 1790–1855*, Cam-
bridge, Mass., 1984. W. R. Albury, 'Experiment and explanation in the physiology of Bichat and
Magendie', *Studies in the History of Biology* (1977), 1, 47–131.

new professional of science, a great part was played by the universities and other institutions of higher learning, especially the French ones which had been reformed as a result of the French Revolution, and the new secular University of Berlin, which was the model for other universities in Prussia and the other German states.[40]

It is now also recognized that in this period a new kind of site dedicated to the production of knowledge about nature first became common and came to be seen as basic to research in the sciences of nature: the laboratory. In chemistry the term 'laboratory' had long been used for the workplace, but now laboratories began to be created in physics, physiology, and later other new sciences such as bacteriology. Professors of the nature-sciences in the German universities were provided by their respective states with their own laboratories and were expected to use them to find out new things about nature. This was the basis of the laboratory research careers of Müller and Liebig, of Helmholz, Virchow and Koch, who were expected to (and did) establish research schools based on teaching laboratories. This pattern of state-provided laboratories was to be envied and emulated by French men of science, by the British and by Americans. In the laboratory research methods were taught in practice, and new generations of experimental researchers into nature were reared. It is now clear that between them they established the laboratory as the main locus of the creation and assessment of most natural knowledge. The laboratory was made the final arbiter of truth about nature.[41]

More controversially, the growing body of work on this period enables it to be characterized as a time when the investigation of nature was changed from a 'godly' to a secular activity. This does not mean that there was a decline in religious belief (although there probably was); the important thing is that religious beliefs became private. It is possible today for scientists to have religious beliefs, but these are supposed to be irrelevant to their science; their religion is

40   For Britain see Jack Morell and Arnold Thackray, *Gentleman of Science: Early Years of the British Association for the Advancement of Science*, Oxford, 1981. Marie Boas Hall, *All Scientists Now: The Royal Society in the Nineteenth Century*, Cambridge, 1984. On France see for instance Dorinda Outram, 'Politics and vocation: French science 1793–1830', *BJHS* (1980), 13, 27–43; Robert Fox, 'Science, the university, and the state in nineteenth century France', in *Professions and the French State 1700–1900* (ed. Gerald' Geison), Philadephia, 1984, 66–145; Robert Fox, 'Scientific enterprise and the patronage of research in France 1800–70', *Minerva* (1973), 11, 442–73. On Germany see R. Steven Turner, 'The growth of professsorial research in Prussia, 1818 to 1848 – causes and context', *Historical Studies in the Physical Sciences* (1971), 3, 137–82; on Berlin, see Elinor S. Shafler, 'Romantic philosophy and the organization at the disciplines: the founding of the Humboldt University of Berlin', In *Romanticism and the Sciences* (ed. Andrew Cunningham and Nicholas Jardine), Cambridge, 1990, 38–54.

41   See the articles in *The Development of the Laboratory: Essays on the Place of Experiment in Industrial Civilization* (ed. Frank A. J. L. James), Basingstoke, 1989; in *The Investigative Enterprise: Experimental Physiology in Nineteenth-Cenury Medicine* (ed. William Coleman and Frederic L. Holmes), Berkeley, Calif., 1988; and in *The Laboratory Revolution in Medicine* (ed. Andrew Cunningham and Perry Williams), Cambridge, 1992.

supposed to be a matter *only* of private belief. As sociologists use the term, 'secularization' means religious institutions giving way to new social institutions in matters of politics, education, social policy and morality.[42] And this kind of change was just what happened in the Age of Revolutions as new political, legal and educational institutions were established across Europe, inspired by the *philosophes*, as with the educational systems of Prussia and Hannover, or imposed by the administrators of Napoleon's Empire. Despite the political changes following from the Bourbon Restoration, most of the new legal, educational and administrative institutions were preserved intact, and even in political terms no country returned to the *ancien régime*. What we are pointing to here is that, paralleling this creation of new secular institutions, there was the creation of new secular disciplines – or the desacralizing of old ones.[43] Epitomizing which is that wonderful if perhaps apocyphal moment when Laplace said to Napoleon – when the world's top physicist said to the world's most powerful man – that he had 'no need of the hypothesis' of God in his account of the Heavenly Mechanism, having effectively, in his *Traité de mécanique céleste* (1799–1825), taken God out of Newton's universe.[44] Coupling this with the observation about the abandoning of the term 'natural philosophy previously used for the study of the natural world as created by the Christian God, our claim can be stated most simply: that 'science' was the new collective name of the new secular discipline for studying the natural world as a secular object, for the discovery of abstract regularities in nature and for the exploitation of natural resources, for acquiring knowledge in secular sense and for material and social improvement.

This is a sketch of these changes in broad brush strokes. Of course, when we come to look in detail, we find that there were local contests for the meaning of the important terms, and the term 'science' in particular was used by many who did not agree that the investigation of the natural world should be a secular activity. This was especially the case in England, which frankly is an exception to this pattern. But we believe that it is the exception which proves the rule, because it was an exception at every level. Although in the eighteenth century England had been the source of inspiration for French and German intellectuals, the defensive reaction of the British to the Franch Revolution meant that England was slow to follow the transformations in the study of nature taking place on the Continent. In England the Church and the

42   See, for example, Tony Bilton et al., *Introductory Sociology*, London, 1981, which defines secularization as 'the process through which religious thinking, practice, and institutions lose their social significance' (p. 531). The crucial word here is 'social'.
43   Owen Chadwick, *The Secularization of the European Mind in the Nineteenth Century*, Cambridge, 1975.
44   On the Laplace story see Roger Hahn, 'Laplace and the mechanistic universe', in *God and Nature: Historical Essays on the Encounter Between Christianity and Science* (ed. David C. Lindberg and Ronald L. Numbers), Berkeley, California, 1986, 256–76.

aristocracy stayed largely in control, and Oxford and Cambridge, the main institutions of learning, were kept avowedly Christian. Exceptional institutions of learning such as the University of London (later University College London), founded in 1826 to represent the radical, God-less and Utilitarian view, were deeply controversial. Even the attempt to found the British Association for the Advancement of Science on the German model of the Gesellschaft Deutscher Naturforscher und Ärzte was almost immediately hijacked by Cambridge dons such as William Whewell and devoted to the pursuit of a Christianized contemplative knowledge of nature.[45] Until mid-century, the most forward-looking amongst young British men went to France and the German states to study. Not until around 1860 did corresponding institutional and intellectual changes happen in England and enable it to catch up with the Continent; hence although we can find individual attempts to model English investigation of the natural world on the Continental pattern, yet still most such investigation by the English remained God-centred – that is, it remained Natural Philosophy – until then.[46]

We are proposing that the origins of science can be located as one aspect of the Age of Revolutions (with England as a partial exception, in that changes there took place later and more gradually than in Continental Europe). These revolutions, as conventionally characterized, are: (1) the French Revolution, beginning in 1789, which was a political revolution, concerned with radically transforming the political organization of society; (2) the industrial revolution, beginning in Britain in the 1770s, a revolution in the means of production, exchange and ownership of the wealth or resources of society; and (3) the post-Kantian intellectual revolution, centred on the German states, a revolution in what one should think and in who should be the intellectual masters of the future. As a result of these simultaneous and linked revolutions, a new middle class was consolidated, wielding the political power, the industrial power, and the intellectual power.

To locate the origins of science in these events nicely conforms to the second historiographic principle outlined above, that the knowledge of any society is an integral product of that society. For it is to be expected that the invention of a new form of intellectual activity should be the product of a major social change, such as the consolidation of a new social class with new power bases.

---

45   J. B. Morrell, 'Brewster and the early British Association for the Advancement of Science', in 'Martyr of Science': Sir David Brewster 1781–1868 (ed. A. D. Morrison-Low and J. R. R. Christie), Edinburgh, 1984, 25–9; Morrell and Thackray, op. cit. (40), especially 63–76, 165–75. [. . .]

46   Frank M. Turner, 'The Victorian conflict between science and religion: a professional dimension', Isis (1978), 69, 356–76. W. H. Brock and R. M. Macleod, 'The scientists' declaration: reflexions on science and belief in the wake of Essays and Reviews 1864–5', BJHS (1976), 9, 39–66. Ruth Barton, 'The X Club: Science, Religion, and Social Change in Victorian England', Ph. D. thesis, University of Pennsylvania, 1976.

Locating the origins of science in the Age of Revolutions is also supported by the first historiographic principle, that the norms and values of science require explanation rather than being truths derived from some transcendent realm. For the emergent middle class, or to be more precise, the emergent *professional* middle class, drew its authority from all three revolutions, the intellectual, the political and the industrial. First they drew aurthority from the intellectual revolution, following which primacy was given to the autonomy of ideas and to the attendant concepts of 'originality' and 'genius', with intellectual achievement being considered to be above the market and something to be judged only by one's peers. Secondly the professional middle class drew authority from the political revolution, as developed into the ideology of liberalism: a philosophy not only of free trade but also of free enquiry (especially criticism of the old powers, Church and aristocracy); a philosophy arguing for the establishment of a new aristocracy of intellectual talent – a 'meritocracy', as we would say today. Thirdly the professional middle class also drew authority from the industrial revolution, for although this benefited the commercial-industrial middle class more directly, the two classes at this time saw their interests as essentially coincident and worked closely together, for example in promoting a vision of progress and prosperity, both social and material, in which the professional middle class's secular knowledge, disseminated by new secular education, would be vital. In these three revolutions can be seen the origin of values and aims – for example, genius, free enquiry, free exchange of ideas, objectivity, disinterestedness – which have been made an integral part of the identity of science.[47]

Finally, the location of the origins of science in the Age of Revolutions is supported by the observation that this period also saw the start of many of the particular stories about the history of science which were handed down to Butterfield's generation, and hence down to our own time. In the same way that long, distinguished national traditions were constructed to support the existence of the new nation states,[48] so a long, distinguished intellectual tradition was now also constructed to support the existence of the new enterprise,

---

47 On intellectual transformation and liberalism, see Raymond Williams, *Culture and Society, 1780–1950*, London, 1958, especially Introduction and chs. 1–2; M. H. Abrams, *The Mirror and the Lamp: Romantic Theory and the Critical Tradition*, London, 1971; Marilyn Butler, *Romantics, Rebels and Reactionaries: English Literature and its Background 1760–1830*, Oxford, 1981; R. Steven Turner, 'The growth of professorial research in Prussia, 1818 to 1848 – causes and context', *Historical Studies in the Physical Sciences* (1971), 3, 137–82; T. W. Heyck, *The Transformation of Intellectual Life in Victorian England*, London, 1982; Christopher Harvie, *The Lights of Liberalism: University Liberals and the Challenge of Democracy 1860–86*, London, 1976; Irene Collins, 'Liberalism and the newspaper press during the French Restoration, 1814–1830', *History* (1961), 46, 17–32; R. Hinton Thomas, *Liberalism, Nationalism and the German Intellectuals 1822–1847*, Cambridge, 1951.

48 See for example Eric Hobsbawm and Terence Ranger (eds), *The Invention of Tradition*, Cambridge, 1984.

science.[49] Building on the twin traditions created in the eighteenth century by the French *philosophes* and the German Romantics, early nineteenth-century men of science located science's origin at one or the other of the periods in which these two groups had located the origin of their own forms of philosophy, namely the seventeenth century and ancient Greece respectively.[50] As they co-opted the people of these two periods to serve their own political and social ambitions, the new men of science naturally recast them in a contemporary mould: the ancient Greeks were represented as having established free enquiry into the natural world, appealing only to natural principles and not to gods, while Galileo, Bacon, Descartes, and Newton were represented as having led the way back to this Greek way of thinking and then gone beyond it by establishing a physicalist, mechanist science, free from the constraints and superstitions of religion.[51] This period, in other words, saw not only the beginning of the history of science, but also the origin of the specific 'Whiggish' or 'present-centred' traditions of the history of science which our own generation of historians have inherited, and from which we have been trying to move away for some twenty-five years. It was also this period which saw the origin of that tradition of political history which Herbert Butterfield first called 'the Whig interpretation'; this was created by Whigs or liberals writing constitutional histories, in which all the causes for which they were currently fighting were presented as having been anticipated long ago and fought for for centuries, so that their values seemed transcendent and eternal, their own victory historically-ordained.[52] Here, surely, is the reason why it has been so natural to adopt critically for the history of science a term originally coined in the context of political history: the triumphalist progressivist 'Whig' traditions in the history of politics and the history of science are twin traditions: they were constructed at the same time, by some of the same people, in the service of the same political interests.[53]

To sum up, historical scholarships over the last twenty years enables us to identify the Age of Revolutions as the period which saw the origin of pretty well every feature which is regarded as essential and definitional of the enter-

---

49   Apparently some people find this claim difficult to countenance; but an argument of this kind was being made as long ago as 1970, by Thomas Kuhn in *The Structure of Scientific Revolutions*, in the chapter on 'The invisibility of revolutions' [. . .]

50   On the *philosophes* see Christie, op. cit. (3), 7–8; I. Bernard Cohen, 'The eighteenth-century origins of the concept of scientific revolution', *Journal of the History of Ideas* (1976), 37, 257–88. [. . .]

51   For an example of 'Whig' history of science being deployed for political ends, see John Tyndall, 'The Belfast address', in *Fragments of Science*, 6th edn, 2 vols, London, 1879, ii, 137–203.

52   See Butterfield, op. cit. (32). See also Peter J. Bowler, *The Invention of Progress: The Victorians and the Past*, Oxford, 1989.

53   For a perfect example of a Whig politician and historian creating a 'Whig' view of the history of the investigation of the natural world, see Lord Macaulay's famous essay on Francis Bacon.

prise of science: its name, its aim (secular as distinct from godly knowledge of the natural world), its values (the 'liberal' values of free enquiry, meritocratic expert government and material progress), and it history.

It is on the strength of this scholarship that we propose that the origins of science should be regarded as being in this period, rather than in the seventeenth century. We must emphasize that we are not arguing for Butterfield's 'origins of modern science' or 'the scientific revolution of the seventeenth century' to be simply moved forward 150 years. Most historians who claim to have identified a fundamental change around 1800 have seen it as akin to the old 'scientific revolution'; indeed, some of them have even called it a 'second scientific revolution'.[54] But to see the change in this way is to stay within the old big picture, based on the old assumptions about the nature of science; thus the change is presented as just a further development in the supposedly universal and eternal human enterprise of investigating nature, just a matter of the same thing ('science') being suddenly done much better, with greater intensity, and being better organized. By contrast, what we are proposing in this paper is something more fundamental: that this period saw the origins of science, in the revised sense: that it saw the *creation* of science's particular and definitive aims, values and practice, not by derivation from some transcendent realm, but as a result of particular human activity in response to the local conditions of material life: an event not of *emergence* but more of *invention*. This term 'invention', which is our preferred term, helps to fix the revised view of science as a contingent, time-specific and culture-specific activity, as only one amongst the many ways-of-knowing which have existed, currently exist, or might exist; and for this reason the phrase which we propose for the fundamental changes which took place in this period is 'the invention of science'.[55] And we can of course now drop the qualifier 'modern' since the term 'science' can only be properly applied in our own time, the modern era.[56] What we are speaking of is therefore not the origins of modern science, but the modern origins of science.

This radical way of interpreting these changes may, we hope, be found useful for teaching and as a heuristic for future research. For the purposes of

54   As far as Cohen (op. cit. (3), 97) could find, the term was first used by Thomas S. Kuhn ('The function of measurement in modern physical science', *Isis* (1961), 52, 161–93, 190). However, the term is now also used in other ways; for example, Stephen G. Brush, *The History of Modern Science: A Guide to the Second Scientific Revolution, 1800–1950*, Ames, Iowa, 1988, uses it to refer to a *late*-nineteenth, early-twentieth century revolution, associated with the breakdown of classical physics. [. . .]

55   There is further discussion of the term 'invention' in Cunningham, op. cit. (38).

56   We think it necessary to make this point explicitly, because the report (in the *Newsletter* of the British Society for the History of Science) of our original conference paper, on which this present paper is based, suggested that we were attempting 'to seek transcendent criteria of "modernity"'. We were not. As we hope is now clear, we are not attempting to seek transcendent criteria of *anything*. In fact, that is exactly what we are arguing against.

this paper it enables us to identify the boundaries of science in time and space and culture: to map out the boundaries of that part of the big picture which we occupy. And as the advocates of the 'scientific revolution' believed their story conformed to the historical evidence and accurately represented what had happened in the past, so we believe 'the invention of science' tells the true story about the origins of science: that those origins are modern.

De-centring the 'Big Picture'

History may be servitude,
History may be freedom. See, now they vanish,
The faces and places, with the self which, as it could, loved them,
To become renewed, transfigured, in another pattern.

T. S. Eliot, *Little Gidding*[57]

On this view, the history of science becomes a relatively short and local matter: extending back less than 250 years, and largely confined to western Europe and America. What are we to say of the rest, of analogous knowledge before this time or in other cultures? In the old big picture, other forms of knowledge appeared as early and more primitive versions of science; thus big picture histories which had chapters on Indian and Chinese 'science' tended to place them after the chapters on tribal societies but before the chapters on ancient Greece. This evolutionary view (in accordance, that is, with evolution as popularly understood rather than as in neo-Darwinian theory),[58] showing growth and progress taking place along a single line, was possible only on the assumption that the pursuit of science was a fundamental part of human nature, a universal enterprise transcendentally derived. For a new big picture, in which science is just one amongst a plurality of ways of knowing the world, other forms of knowledge must be allowed to appear on their own terms, instead of being measured against a scientific framework. Science will appear only as the native knowledge-form of our own culture, not in a central or special place. What is required, we might say, is a kind of 'de-centring'.

'De-centring' is a psychologists' term, deriving from the theories of Mead and Piaget, according to which very young children's understanding of the world is at first entirely 'centred' on themselves; 'de-centring' thus describes the process by which children come to realize that external objects have per-

57 Excerpt from 'Little Gidding' in *Four Quartets*, copyright 1943 by T. S. Eliot and renewed 1971 by Esme Valerie Eliot, reprinted by permission of Harcourt Brace &Company, and Faber and Faber Ltd.
58 Strict neo-Darwinian theory, based on natural selection, implies that evolutionary lines branch like the twigs of a bush: a non-linear and non-hierarchical view of evolution (see, for example, Stephen Jay Gould, *Wonderful Life: The Burgess Shale and the Nature of History*, London, 1989, 27–45), Nevertheless, the popular view of evolution is one of a linear, progressive ascent. [. . .]

manence, that other people can have different visual perceptions of the same scene, and that other people can have different knowledge, interests, feelings and so on.[59] In a more general sense, de-centring is something which we continue to do repeatedly throughout our adult lives, as we identify yet another aspect of our own egotism, and realize that something which we thought was universal is actually peculiar to ourselves, or our group, our class, our nation, or our culture. To see science as a contingent and recently-invented activity is to make such a de-centring, and to acknowledge that things about our primary way-of-knowing which we once thought were universal are actually specific to our modern capitalist, industrial world.

This kind of de-centring is already beginning to be made in the history of religion, which until recently, in western European countries, was firmly centred on Christianity. This was quite explicitly and unapologetically the case in traditional providential Church History, in which other religions figured only in relation to Christianity; most spectacularly, Judaism was relegated to the status of a precursor. But even in the more liberal viewpoint, in which different faiths are thought of as moving towards the same end, or as being different parts of the same truth, the Christian-centredness remains, because that element of sameness is conceived in Christian terms; features of Christianity, such a monotheism or belief in the existence of an immortal soul, are taken to be constitutive of religion in general.[60] But during the course of the last two decades, a few writers, mainly working in comparative religion or (in Britain) the teaching of Religious Education, have begun to build a view of religion and its history that is not centred on Christianity but which tries to treat at the very least all the major world faiths symmetrically.[61]

The problem which we face in the History of Science is essentially the same. We too have the legacy of a big picture in which historical events have been interpreted as leading towards our own culture, providentially guided from a transcendent realm – in this case the transcendent element being objective truth, goodness or human nature, rather than the Christian God. There are parallels between the way in which the first practitioners of Christianity attempted to erase the separate identity of Judaism and the way in which the first practitioners of science attempted to erase the separate identity of natural philosophy; in both cases, the older texts were taken over by the new practice and reinterpreted as marking early stages in its development: a reinterpretation that was validated only by assuming the transcendent, eternal and uni-

59   See, for example, George Herbert Mead, Mind, Self, and Society (ed. Charles W. Morris), Chicago, 1934; Jean Piaget, The Child's Conception of Reality (tr. Margaret Cook), London, 1955.
60   This Christian-centredness is revealed, for example, in the common statement that 'different religions are worshipping God in different ways'.
61   Ninian Smart, The Religious Experience of Mankind, New York, 1969, London, 1971; Don Cupitt, Taking Leave of God, London, 1980; Wilfred Cantwell Smith, Towards a World Theology: Faith and the Comparative History of Religion, London, 1981; Keith Ward, A Vision to Pursue: Beyond the Crisis in Christianity, London, 1991; Jean Holm, The Study of Religions, London, 1977. [. . .]

versal nature of Christianity, in the one case, and of objective scientific knowledge, in the other. Just as we need a big picture of religion which does justice to the separate identity of Judaism and other non-Christian faiths, so we need a big picture of the history of science which does justice to the separate identity of natural philosophy and other past and present ways of knowing the natural world.

It is not too difficult to imagine a new big picture which is de-centred from our own position along the axis of time; which grants a separate identity to, for example, ancient Greek philosophy, medieval natural philosophy, and the modified forms of natural philosophy developed in the early modern period; historiography for these periods already exists which points clearly in this direction. De-centring from our own position in space, however, is a more difficult step to take; we are much less aware of the existence, let alone the separate identity, of ways of knowing the natural world outside our own culture, or even of those which we claim as ancestors. One way of seeing the necessity for such a de-centring is to look at a map of the world, preferably in the Peters projection,[62] and consider that almost all the material with which the History of Science discipline has been concerned comes from a tiny geographical area, about the same size as Zaire or the Sudan, and considerably smaller than Brazil. The only thing that is unusual about the countries in this area, apart from the fact that they are where we live, is that it was these countries which rose to world-domination during the nineteenth century, through the formation of overseas empires. It was only this historical accident that has meant that what began as their own native culture – by that time including that recent invention, science – has now become world-culture.[63] A spatially or rather geographically de-centred big picture would treat all native knowledge-forms with perfect symmetry.

But even de-centred temporally and spatially in this way, our big picture might still retain its present almost exclusive focus on cognitive knowledge, following our culture's peculiar elevation of theoretical knowledge to a higher

---

62   The Peters projection is claimed to provide a more accurate representation of the relative size of the Earth's major land areas than the more-familiar Mercator projection, which exaggerates the area of countries further away from the Equator (e.g. Europe). See Arno Peters, *Der Europazentrische Charaktèr unseres geographischen Weltbildes und seine Überwindung*, Dortmund, 1976.
63   Much of the literature on the exporting of Western knowledge is based on the assumption that the process has been more-or-less successful; for example, Peter Buck, *American Science and Modern China, 1876–1936*, Cambridge, 1980; James R. Bartholomew, *The Formation of Science in Japan*, New Haven and London, 1989. But interestingly, some recent works have emphasized the *difference* of the non-Western traditions, hence questioning how completely Western science retained its identity when transplanted. See for example Arnab Rai Choudhuri, 'Practising Western science outside the West: personal observations on the Indian scene', *Social Studies of Science* (1985), 15, 475–505; *Science, Hegemony and Violence: A Requiem for Modernity* (ed. Ashis Nandy), Oxford, 1988; Masao Watanabe, *The Japanese and Western Science* (tr. Otto Theodor Benfey), Philadelphia, 1991, original Japanese edn 1976.

status than practical knowledge. To de-centre further, then, we would need a big picture which dealt not only with cognitive knowledge, the knowledge of fact, but also with practical knowledge, the knowledge of skill: not only with *know-of*, but also with *know-how*. To some extent, this de-centring is already being made, particularly with the work on skills now being done in the Sociology of Scientific Knowledge, and with the growth of the History of Technology as a discipline. But even this new work retains a very strong centre, being concerned with subjects such as measurement and calculation, power sources and industrial processes: those things that have importance principally in industrial societies, and principally for men within industrial societies. Feminist and environmentalist analysis points to some further de-centring which is needed: away from the rather specialized technical achievements, such as travelling very fast, and building tall buildings, and killing people in large numbers; and towards more basic and more generally-appreciated technical achievements, such as making sure that people have enough to eat, and clothing to keep them warm, and a place to sleep, and comfort and healing when they get sick. To incorporate such know-how within a big picture should surely be a feasible goal.[64]

De-centring could be taken still further. Cognitive knowledge – the knowledge of fact, and technical knowledge – the knowledge of skill, are only two aspects of the knowledge which each one of us possesses. There is also what we might call relational knowledge: the knowledge of acquaintance, the knowledge by which each of us relates to another, person to person, and in small groups, and in larger social and political units. There is also moral knowledge: the knowledge of value, by which each of us judges what is right and wrong. Would it be possible to write a history of human knowledge which, instead of following the positivist legacy of elevating cognitive knowledge (and only one kind of cognitive knowledge, at that) above all the rest, would treat fact, skill, human relations and morals with perfect symmetry?[65]

We do not apologize for raising such questions, even if we cannot yet see how they might be followed out. A big picture, as we understand it, is not just a summary of research results or a general theory of history, but a vision of the world and our place in it; and if such a vision is to remain alive, to grow and to change, then it is necessary for its limits to be constantly pushed beyond what is currently imaginable. Of course, all of us will probably continue to remain specialists, more or less; even when teaching general courses, for some

---

64   This paragraph was inspired by Ursula K. Le Guin, 'The carrier bag theory of fiction', in her *Dancing at the Edge of the World: Thoughts on Words, Women, Places*, London, 1989, 165–70. See also Joan Rothschild (ed.), *Machina ex Dea: Feminist Perspectives on Technology*, New York, 1983. An interesting recent attempt at a 'big picture' history from an ecological perspective – i.e. of humanity's changing relationship to its environment, through the development of agriculture and industry – is Clive Ponting, *A Green History of the World*, London, 1991.
65   This paragraph was inspired by Hilary Rose, op. cit. (21).

considerable time to come we are likely to continue to specialize in the dominant Western traditions of knowledge – philosophy, natural philosophy, and
science. But since these specialisms derive their meaning and identity from the
big picture in which, explicitly or implicitly, they are placed, we should do our
best to ensure that the big picture we use is one in which we believe and which
is appropriate to our time and place and culture. We owe this to our students,
to our public, and to ourselves.

Glossary

The following is a glossary of terms and people not explained in the text. See also Wilbur Applebaum (ed.), *The Encyclopedia of the Scientific Revolution* (New York and London: Garland 2000).

Accademia dei Lincei   the Academy of the Lynxes; society founded by Prince Cesi in 1603 to support mathematics and natural philosophy.

Aristarchus of Samos (ca. 310–230 BCE)   ancient Greek astronomer who suggested that the earth rotated on its axis and revolved around the Sun.

Averroës (1126–98)   Spanish Arab philosopher whose reconciliation of philosophy and theology was controversial throughout the Middle Ages.

Bacon, Francis (1561–1626)   English lawyer and courtier who sought to reform natural philosophy to make it practical and useful.

Basso, Sebastian (ca. 1560–1625)   French physician and natural philosopher who propounded atomism while strongly criticizing Aristotelian natural philosophy.

Borelli, Giovanni Alfonso (1608–79)   Italian mathematician and physiologist who sought to explain animal motion according to mechanics.

Boyle, Robert (1627–91)   Anglo-Irish aristocrat and experimental philosopher, active in the Royal Society.

Brahe, Tycho (1546–1601)   Danish nobleman who raised astronomical observation to new levels of accuracy.

Camões, Luís de (ca. 1524–80) Portuguese poet and voyager whose epic *Lusiads* describes Vasco da Gama's discovery of the sea route to India.

Cesi, Frederico (1585–1630) Italian nobleman who founded the Accademia dei Lincei.

Colbert, Jean-Baptiste (1619–83) French minister of state under Louis XIV, founded the Royal Academy of Science in Paris in 1666.

Collegio Romano the Jesuits' flagship university, located in Rome.

Copernicus, Nicholas (1473–1543) Polish Church administrator and doctor, published a heliocentric cosmology in 1543.

Cusanus, Nicolaus [Nicholas of Cusa] (1401–64) German cardinal, mathematician, philosopher, and theologian; a model Renaissance man.

Duhem, Pierre (1861–1916) French physicist and historian of medieval science.

epistemology the study of the nature and limits of knowledge.

Euclid (fl. ca. 300 BCE) Greek geometer, best known for his *Elements*, the standard geometry text into the seventeenth century.

Galen (CE 129–ca. 216) Greek physician whose humoral theories of disease dominated physiology into the seventeenth century.

Galilei, Galileo (1564–1642) Italian mathematicians and natural philosopher, made important astronomical discoveries with the telescope, attempted to reformulate the study of mechanics.

Gassendi, Pierre (1592–1655) French priest, mathematician and natural philosopher who revived ancient atomism.

geocentric having the earth as the center of motion.

geoheliocentric having two centers of motion, the earth and the Sun.

Gilbert, William (1544–1603) English natural philosopher and physician who pioneered the study of magnetism in his *De Magnete* of 1600.

Gresham College London institution founded under the will of Sir Thomas Gresham in 1596 to teach useful disciplines. It became the home of the Royal Society.

heliocentric having the Sun as the center of motion.

Henry the Navigator (1394–1460)   Portuguese prince, supported voyages of discovery.

Hermeticism   general term for esoteric and occult disciplines, supposedly based on the works of the ancient magus, Hermes Trismegistus.

Hobbes, Thomas (1588–1679)   English philosopher who approved of mechanical natural philosophy while rejecting the experimental method.

Hooke, Robert (1635–1703)   English mathematician and experimental philosopher, curator of experiments at the Royal Society.

Huguenot   French Protestant minority.

humanism   movement that sought to revive ancient literature, ethics, and history and employ them in education and public affairs.

Huxley, T. H. (1825–95)   English natural historian and biologist, nicknamed "Darwin's Bulldog" for his public support of the theory of evolution.

Huygens, Christiaan (1629–95)   Dutch mathematician and natural philosopher.

impetus theory   a medieval theory that claimed bodies move because a kind of motive force is impressed in them, but slow and stop as it wears off.

intellectualism   the philosophical position that God is constrained by a necessary and rational natural order.

Kepler, Johannes (1571–1630)   German mathematician and Copernican astronomer, best known for his discovery that planetary orbits are elliptical.

Leibniz, Gottfried Wilhelm (1646–1716)   German philosopher and mathematician, developed the calculus independently of Newton.

Linus, Francis (1595–1675)   English Jesuit who disputed Robert Boyle's explanation of the phenomena produced by the air-pump.

Molière (1622–73)   French playwright who satirized contemporary society.

More, Henry (1614–87)   English poet and philosopher who was a member of the group known as the Oxford Platonists.

Newton, Isaac (1642–1727)   English mathematician and natural philosopher; among his remarkable achievements he developed the theory of universal gravity, classical mechanics, and the calculus.

Nunes, Pedro (1502–78)   Portuguese cartographer, geographer, and professor of mathematics.

Oldenburg, Henry (ca. 1618–77)　German living in England; as first secretary of the Royal Society he corresponded with scholars across Europe.

Organum　collective term for Aristotle's works of logic meaning "tool." Francis Bacon attempted to replace it with his *New Organum*.

Paracelsus (1493–1541)　German-Swiss alchemist and physician who criticized ancient learning, in particular Galen, and advocated chemical medicines.

Pascal, Blaise (1623–62)　French mathematician and philosopher, known for his work on probability and his mercury tube experiments.

Policiano, Angelo [Politian] (1454–1594)　Florentine humanist, member of Platonic academy, outstanding poet and classical scholar.

probabiliorism　moral teaching that one must follow the more probable of two or more possible opinions or actions.

Proclus (ca. CE 410–485)　Greek philosopher who sought to defend paganism against Christianity with Neoplatonic philosophy.

Ptolemy (fl. CE 150)　Greek astronomer who lived in Alexandria, Egypt. His *Almagest* was the standard work of technical astronomy until the seventeenth century.

Ramus, Peter [Pierre de la Ramée] (1515–72)　French educational reformer who sought a new curriculum based on logic and rhetoric.

Restoration　the period following the reestablishment of the monarchy in England under Charles II in 1660.

Scholasticism　approach to philosophy widely adopted at universities, based largely upon the works of Aristotle and his medieval interpreters.

Seneca (ca. 4 BCE–CE 65)　Roman philosopher who wrote works of practical ethics derived from Stoicism.

Sennert, Daniel (1572–1637)　German chemist, physician, and natural philosopher who contributed to the revival of atomism.

Sorbonne　the theological faculty of the University of Paris.

Sprat, Thomas (1635–1713)　English clergyman, wrote *History of the Royal Society* in 1667, an outline of the Society's methods and aims.

teleology　explanation from the perspective of purpose or final cause.

Tempier, Étienne (Stephen)　Bishop of Paris, published list of forbidden propositions in 1277.

Torricelli, Evangelista (1608–47)   student of Galileo, mathematician and inventor of the mercury tube barometer.

Tridentine   relating to the Council of Trent, a general council of the Catholic Church that met to establish Catholic doctrine in response to the Protestant Reformation.

trireme   an ancient Greek galley.

via moderna   also known as nominalism; a medieval school of philosophy critical of Aristotle.

vitriol   chemical substance now known as sulfuric acid whose color varies with its metallic composition.

Vives, Luis (1492–1540)   Spanish humanist, who advocated education based on the study of natural things.

voluntarism   the philosophical position that God can do whatever he wills even if this appears logically or naturally impossible.

Wittgenstein, Ludwig (1889–1951)   Austrian philosopher whose philosophy of language has been extremely influential in the recent cultural and social approach to the history and sociology of science.

INDEX